RUDIMENTS OF PLANE AFFINE GEOMETRY

P. SCHERK AND R. LINGENBERG

P. SCHERK is Professor of Mathematics at the University of Toronto. R. LINGENBERG is Professor of Mathematics at the Universität Karlsruhe, Germany, and the author of three books on linear algebra and geometry.

In recent decades research into the foundations of geometry has completely transformed the field and created a need for new methods of presenting geometry at the university level. *Rudiments of Plane Affine Geometry* develops rigorously and clearly one geometric theory accessible to the reader with no previous experience. It reflects the spirit and displays some of the basic ideas of modern geometric axiomatics. The volume is intended for undergraduates with a modest knowledge of algebra and linear algebra and provides a sound introduction to deductive geometry.

MATHEMATICAL EXPOSITIONS

Editorial Board

H.S.M. COXETER, G.F.D. DUFF, D.A.S. FRASER,
G. de B. ROBINSON (Secretary), P.G. ROONEY

Volumes Published

1 *The Foundations of Geometry*   G. DE B. ROBINSON
2 *Non-Euclidean Geometry*   H.S.M. COXETER
3 *The Theory of Potential and Spherical Harmonics*   W.J. STERNBERG and T.L. SMITH
4 *The Variational Principles of Mechanics*   CORNELIUS LANCZOS
5 *Tensor Calculus*   J.L. SYNGE and A.E. SCHILD
6 *The Theory of Functions of a Real Variable*   R.L. JEFFERY
7 *General Topology*   WACLAW SIERPINSKI
  (translated by C. CECILIA KRIEGER) (out of print)
8 *Bernstein Polynomials*   G.G. LORENTZ (out of print)
9 *Partial Differential Equations*   G.F.D. DUFF
10 *Variational Methods for Eigenvalue Problems*   S.H. GOULD
11 *Differential Geometry*   ERWIN KREYSZIG (out of print)
12 *Representation Theory of the Symmetric Group*   G. DE B. ROBINSON
13 *Geometry of Complex Numbers*   HANS SCHWERDTFEGER
14 *Rings and Radicals*   N.J. DIVINSKY
15 *Connectivity in Graphs*   W.T. TUTTE
16 *Introduction to Differential Geometry and Riemannian Geometry*   ERWIN KREYSZIG
17 *Mathematical Theory of Dislocations and Fracture*   R.W. LARDNER
18 *n-gons*   FRIEDRICH BACHMANN and ECKART SCHMIDT
   (translated by CYRIL W.L. GARNER)
19 *Weighing Evidence in Language and Literature: A Statistical Approach*
   BARRON BRAINERD
20 *Rudiments of Plane Affine Geometry*   P. SCHERK and R. LINGENBERG

MATHEMATICAL EXPOSITIONS NO. 20

Rudiments of plane affine geometry

P. SCHERK and R. LINGENBERG

UNIVERSITY OF TORONTO PRESS
Toronto and Buffalo

© University of Toronto Press 1975
Toronto and Buffalo
Printed in Great Britain

LIBRARY OF CONGRESS CATALOGING IN PUBLICATION DATA

Scherk, Peter
Rudiments of plane affine geometry

(Mathematical expositions; no. 20)
Bibliography: p.
Includes index.
1. Geometry, Affine.　I. Lingenberg, Rolf, 1929–　joint author.
II. Title.　III. Series.
QA477.S33　　516′.4　　75-11705
ISBN 0-8020-2151-4　　CN ISSN 0076-5333
AMS 1970 subject classifications
50–A99, 50–D05

IN MEMORY OF HANS HEILBRONN

Contents

PREFACE ix

1
AFFINE PLANES 3
1.1 The axioms 3
1.2 Examples 7
1.3 A co-ordinate plane 8
1.4 Finite planes 10
Exercises 13

2
COLLINEATIONS 14
2.1 Bijections 14
2.2 Collineations 15
2.3 Fixed elements 19
2.4 Homotheties 21
2.5 Translations 23
2.6 Dilatations 26
2.7 Axial affinities 28
Exercises 30

3
TRANSLATION PLANES 32
3.1 Linear transitivity 32
3.2 The configurations of translation planes 32
3.3 The prime kernel of a translation plane 44
Exercises 49

4
DESARGUESIAN PLANES 52
4.1 The dilatation groups $D(O)$ 52
4.2 The shear theorem 57

4.3 The linear transitivity of the groups $A(a)$ — 61
Exercises — 64

5
PAPPUS PLANES — 66

Appendix — 69
Exercises — 71

6
CO-ORDINATES IN DESARGUESIAN PLANES — 72

6.1 Co-ordinate planes — 72
6.2 Co-ordinates in desarguesian planes — 74
6.3 The fundamental theorem of affine geometry — 80
Exercises — 84

7
THE PROJECTIVE CLOSURE OF AN AFFINE PLANE — 85

7.1 Motivation — 85
7.2 The axioms of a projective plane — 86
7.3 The duality principle — 88
7.4 Collineations — 90
7.5 The canonical duality — 95
7.6 Affine restrictions of collineations of projective planes — 96
Exercises — 99

8
DESARGUESIAN PROJECTIVE PLANES — 101

8.1 Projective and affine desarguesian planes — 101
8.2 The projective theorem of Desargues — 102
8.3 Projective Pappus planes — 104
Exercises — 105

APPENDIX — 107

A.1 Groups — 107
A.2 Skew fields — 108
A.3 Right vector spaces — 108

REFERENCES — 111

INDEX — 113

Preface

There is fairly general agreement that the traditional presentation of euclidean geometry at the colleges (and high schools) on this continent is an anachronism. We have to look for alternatives if we wish to keep geometry alive.

Affine geometry is both modern and simple. It is closely related to euclidean geometry. We can introduce it by stripping euclidean geometry of its topological and metric features. The beginnings of the resulting theory are simple enough. It is enriched gradually until the transition to affine (and euclidean) analytic geometry, which is more or less hidden in today's linear algebra, can readily be made.

This book is suitable for a one-term course for undergraduates with a modest knowledge of algebra and linear algebra. Its detailed presentation is the result of our intention to make it suitable for the beginner and the non-specialist. Some of the more difficult sections and problems are marked by asterisks and can be omitted if necessary. Some instructors may also wish to skip the proof of Theorem 3 of Chapter 2. For the convenience of the reader some algebraic concepts are collected in an appendix. It will not be referred to in the text.

The literature is extensive and recent. Very little in this book could have been written thirty years ago. The interested reader will find references at the end of the text.

In conclusion the authors wish to thank Dr E. Ellers for his careful reading of their manuscript and his criticisms.

P. SCHERK and R. LINGENBERG
Toronto and Karlsruhe
March 1974

RUDIMENTS OF PLANE AFFINE GEOMETRY

1
Affine planes

1.1
THE AXIOMS

Euclidean geometry as it is taught in our high schools is a highly sophisticated theory. It combines three classes of concepts: (i) linear concepts such as 'point,' 'straight line,' 'incidence,' 'parallelism,' etc.; (ii) metric concepts such as 'length' and 'angle' as well as 'orthogonality' and 'congruence'; (iii) topological concepts such as 'betweenness.'

We arrive at *affine geometry* by dropping both the metric and the topological concepts. Thus the purpose of this book is the discussion of those parts of euclidean geometry which deal with linear concepts only. Our treatment will have to avoid metric or topological reasoning.

We shall start with some natural assumptions which are satisfied in euclidean geometry. They will be our axioms. (Every mathematical theory needs, in addition to some basic undefined concepts and relations, some assumptions ['axioms'] on them which we agree to accept. A 'proof' reduces other statements ['theorems'] to these axioms.*)

Let **P** and **L** denote two non-void disjoint sets. We call the elements of **P** *points* and those of **L** (straight) *lines*† and denote the former by capital letters $P, Q, A, B, \ldots$ and the latter by small ones $g, h, l, \ldots$. In addition, we are given a *relation* I called *incidence* between certain points and lines. Thus I is a subset of the set of all the pairs (P, l) of points and lines. If the pair (P, l) belongs to this subset, we write

$$P \mathrm{I} l \quad \text{or} \quad l \mathrm{I} P$$

and say 'P and l are incident' or 'P lies on l' or 'l passes through P.' If P and l are not incident, we write

$$P \not{\mathrm{I}} l.$$

---

* The reader should be warned though that some of the 'planes' which satisfy these axioms will look rather different from the euclidean plane.
† In this text, 'line' will be synonymous with 'straight line.' Note that 'line' means 'curve' in differential geometry.

If l is incident with two points P and Q, i.e. if

$$l \text{ I } P, \quad l \text{ I } Q, \tag{1}$$

we say l *connects* P and Q.

If the point P is incident with the lines l and h, i.e. if

$$P \text{ I } l, \quad P \text{ I } h, \tag{2}$$

we say l and h have the point P in common.

The lines l and h are said to be *parallel*,

$$l \parallel h,$$

if either $l = h$ or l and h have no point in common. Thus two distinct lines l and h are parallel if (2) has no solution P.

If l and h are not parallel, we write

$$l \nparallel h.$$

We state

AXIOM A1 *If $P \neq Q$ there exists one and only one line connecting P and Q.*

We denote this line by $[P, Q]$. Thus $l = [P, Q]$ is equivalent to formula (1) if $P \neq Q$.

THEOREM 1 *Two non-parallel lines l and h have exactly one point in common.*

PROOF We have $l \nparallel h$ if and only if $l \neq h$ and l and h have at least one point in common. Thus we have to prove that l and h have not more than one point in common.

Suppose l and h have the distinct points P and Q in common. Then P and Q would be connected by the two distinct lines l and h. This contradicts Axiom A1. □

If $l \nparallel h$, we say that l and h *intersect* and call the common point the *intersection* of l and h. It will be denoted by

$$[l, h] = [h, l].$$

Our next assumption is the parallel axiom of euclidean geometry.

AXIOM A2 *Given l and P, $P \not\hspace{-2pt}\text{I } l$, there exists one and only one line through P parallel to l.*

If $P \text{ I } l$, there is at least one line through P parallel to l, viz. l itself. If $h \neq l$ is another line through P, then h and l are both incident with P. Hence $h \nparallel l$. Thus we can improve Axiom A2 slightly:

1.1 THE AXIOMS

THEOREM 2 *To every line l and every point P there is exactly one line through P parallel to l.*

THEOREM 3 *Parallelism is an equivalence relation.*

We remember that a relation is called an *equivalence relation* if it is reflexive, symmetric, and transitive.

(i) *Reflexivity* We have to show that every line is parallel to itself. This follows from our definition.

(ii) *Symmetry* To show that $l \parallel h$ implies $h \parallel l$. This is true if $l = h$. Let $l \neq h$; $l \parallel h$. Then l and h have no point in common. Thus $h \parallel l$. (Or more briefly: our definition of $l \parallel h$ is symmetric in l and h.)

(iii) *Transitivity* Let $l \parallel g$, $g \parallel h$. We have to show that $l \parallel h$. This is certainly true if l and h have no common point. Suppose then that l and h are both incident with P. By our assumptions and by part (ii) of our proof we have

$l \parallel g, l \, I \, P$ and $h \parallel g, h \, I \, P$.

Thus both lines l and h pass through P and are parallel to g. Hence by Axiom A2, $l = h$ and therefore $l \parallel h$. $\square$

Let Π_l denote the set of the lines parallel to l. Thus

$$\Pi_l = \{h \mid h \parallel l\}$$

and

$$l \in \Pi_l. \tag{3}$$

We call Π_l a *parallel pencil*. Our next goal is

THEOREM 4

(i) *Every line belongs to one and only one parallel pencil.*

(ii) *Two lines l and h are parallel if and only if they belong to the same parallel pencil.*

PROOF

(i) Let l be any line. By (3) l belongs to the parallel pencil Π_l. Suppose the parallel pencil Π_h also contains l. We have to show that $\Pi_h = \Pi_l$.

Let $g \in \Pi_h$. Thus $g \parallel h$. Since $l \in \Pi_h$, we have $l \parallel h$. Hence by Theorem 3, $g \parallel l$ or $g \in \Pi_l$. Since this applies to every $g \in \Pi_h$, we obtain

$$\Pi_h \subset \Pi_l. \tag{4}$$

Conversely, assume $g \in \Pi_l$. Thus $g \parallel l$. Since $l \parallel h$, this yields $g \parallel h$ or $g \in \Pi_h$ and therefore

$$\Pi_l \subset \Pi_h. \tag{5}$$

Combining (4) and (5), we obtain $\Pi_h = \Pi_l$.

(ii) If $n \parallel l$, then h and l both belong to Π_l. Conversely, if h and l both lie in Π_g, then $h \parallel g$ and $l \parallel g$, and therefore $h \parallel l$.□

THEOREM 5 *Let $l_1 \parallel l_2$, $l_1 \nparallel h$. Then $l_2 \nparallel h$. The points $[l_1, h]$ and $[l_2, h]$ are equal if and only if $l_1 = l_2$.*

PROOF Suppose that $l_2 \parallel h$. Then $l_1 \parallel l_2$ implies $l_1 \parallel h$ by Theorem 3. Thus $l_2 \nparallel h$ and the points $[l_1, h]$ and $[l_2, h]$ exist by Theorem 1.

Obviously, $l_1 = l_2$ implies $[l_1, h] = [l_2, h]$. Conversely, if $l_1 \neq l_2$, the points $[l_1, h]$ and $[l_2, h]$ cannot be equal since they are incident with distinct parallel lines.□

We call three points *collinear* if there is a line that all three of them are incident with. One third assumption reads:

AXIOM A3 *There are three non-collinear points.*

A triplet

$$\mathfrak{A} = (\mathbf{P}, \mathbf{L}, \mathtt{I})$$

satisfying the axioms A1–A3 is called an *affine plane*.

Note that **P** is not void by Axiom A3 and that Axiom A1 then ensures the existence of lines.

We call a triplet A, B, C of non-collinear points a *triangle ABC* and the lines

$$a = [B, C], \qquad b = [C, A], \qquad c = [A, B] \tag{6}$$

the *sides* of the triangle.

THEOREM 6 *No two sides of a triangle are parallel.*

PROOF See Exercise 6.

The set of all the lines incident with a given point P is called the *pencil of the lines through P*.

THEOREM 7
(i) *Every parallel pencil contains at least two lines.*
(ii) *Every line is incident with at least two points.*
(iii) *For every P, the pencil of the lines through P contains at least three lines.*

PROOF By Axiom A3, there exists a triangle ABC with the sides a, b, c.
(i) Let Π_l denote a parallel pencil. Then at least one of the points A, B, C, say A, is not incident with l. The line through A parallel to l belongs to Π_l and is distinct from l.
(ii) Given the line l, Theorems 3 and 6 imply that at least one of the sides of ABC, say a, is not parallel to l. We have already shown that Π_a contains a line

$a' \neq a$. Since $a' \parallel a$ and $a \nparallel l$, Theorem 5 yields $a' \nparallel l$ and $[a, l] \neq [a', l]$. Thus l is incident with the distinct points $[a, l]$ and $[a', l]$.
(iii) Let P be arbitrary. Let a', b', c' denote the lines through P parallel to a, b, c, respectively. By Theorems 6 and 4, the three parallel pencils Π_a, Π_b, Π_c are mutually disjoint. Hence the three lines a', b', c' through P are distinct. $\square$

THEOREM 8 *Two lines are non-parallel if and only if they have exactly one point in common.*

PROOF By Theorem 1 two non-parallel lines l and h have exactly one point in common. Suppose then $l \parallel h$. If $l \neq h$, l and h have no point in common. If $l = h$, there are, by Theorem 7(ii), at least two points on this line. $\square$

1.2
EXAMPLES

To familiarize ourselves with the three axioms, we first mention three somewhat intuitive examples from euclidean geometry.
(i) The plane euclidean geometry satisfies our three axioms.
(ii) The geometry on the euclidean line satisfies A1 and A2 but not A3.
(iii) **P** consists of the points of euclidean three-space, **L** of its lines. 'Incidence' means ordinary incidence. Thus two skew lines are 'parallel' and A2 does not hold. However, A1 and A3 remain in force.

The next examples are rigorous. They will demonstrate the independence of the three axioms A1–A3; cf. Exercises 4–6 and Section 1.4.
(iv) Let **P** be any non-void set and **L** a set consisting of one element $l \notin \mathbf{P}$. The relation I consists of all the pairs (P, l) with $P \in \mathbf{P}$. Thus $P \text{ I } l$ for all $P \in \mathbf{P}$. Then the triplet (**P**, **L**, I) satisfies A1 and A2 but not A3; cf. Example (ii).
(v) Let $\mathbf{P} = \{A, B, C\}$ and $\mathbf{L} = \{a, b, c\}$ denote any two disjoint sets of three elements each. The incidence relation consists of the pairs.

(A, b), (A, c), (B, a), (B, c), (C, a), (C, b).

Thus

$A \text{ I } b, c,$ $B \text{ I } a, c,$ $C \text{ I } a, b;$

cf. Figure 1.1. Then the triplet (**P**, **L**, I) satisfies A1 and A3 but not A2 because no two distinct lines are parallel.
(vi) Let $\mathbf{P} = \{A, B, C\}$ be an arbitrary set of three elements and let $\mathbf{L} = \{a, d\}$ be any set of two other elements. Furthermore the relation I consists of the pairs (A, d), (B, a), (C, a). Thus

$A \text{ I } d,$ $B \text{ I } a,$ $C \text{ I } a;$

cf. Figure 1.2. Then the triplet (**P**, **L**, I) satisfies A2 and A3 but not A1.

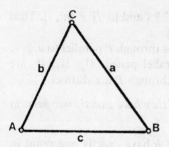

Figure 1.1

In the next two sections we shall discuss some triplets (**P**, **L**, **I**) which satisfy the axioms A1–A3, thus obtaining rigorous examples of affine planes.

1.3
A CO-ORDINATE PLANE

The following example illustrates an important method of constructing affine planes.

Let $\mathbb{R}$ denote the field of the real numbers. Put

$$\mathbf{P} = \mathbb{R} \times \mathbb{R} = \{(x, y) \mid x, y \in \mathbb{R}\}.$$

Thus the 'points' are pairs (x, y) of real numbers. The 'lines,' i.e. the elements of **L**, are certain point sets, viz.:

(i) for every $c \in \mathbb{R}$ we take the point set

$$l = l(c) = \{(x, y) \mid x = c\}. \tag{7}$$

(ii) for every pair of elements m, d of $\mathbb{R}$, the point set

$$l = l(m, d) = \{(x, y) \mid y = mx+d\} \tag{8}$$

shall be a line.

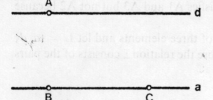

Figure 1.2

1.3 A CO-ORDINATE PLANE

The incidence relation I consists of all the pairs (P, l) such that the point P is an element of the point set l. Thus

$$P \,\mathrm{I}\, l \Leftrightarrow P \in l. \tag{9}$$

Before we prove that the triplet $\mathfrak{A} = (\mathbf{P}, \mathbf{L}, \mathrm{I})$ is an affine plane we investigate parallelism. By definition two lines are parallel if they are either identical or disjoint.

(α) Obviously, any two lines $l(c)$ and $l(c)'$ are parallel.
(β) No two lines $l = l(m, d)$ and $l' = l(c)$ are parallel.
Evidently $l \neq l'$ and the point $(c, mc+d)$ lies on both l and l'.
(γ) $l(m, d) \parallel l(m', d')$ if and only if $m = m'$.
Let $l = l(m, d)$ and $l' = l(m', d')$. Assume first $m = m'$.
If $d = d'$, then $l = l'$; thus $l \parallel l'$. If $d \neq d'$, l and l' have no common point (x, y); for it would satisfy

$$mx+d = y = m'x+d' = mx+d'.$$

Thus $d = d'$, contrary to our assumptions.

Conversely, let $l \parallel l'$. If $m \neq m'$, the equation $mx+d = m'x+d'$ has a solution x and the point $(x, mx+d)$ would be incident with both l and l'. Since $l \parallel l'$, this implies $l = l'$ and $m = m'$; contradiction.

We now verify the axioms A1–A3.

To Axiom A1 Let $P = (a, b)$ and $P' = (a', b')$ denote two distinct points in $\mathfrak{A}$. If $a = a'$, then $l(a)$ connects P and P'.

Suppose there is another line connecting P and P'. Then it must be a line $l(m, d)$. Thus $b = ma+d = ma'+d = b'$ or $P = P'$; contradiction.

For $a \neq a'$ there is one and only one m such that $m(a-a') = b-b'$. Put $d = b-ma$. Then both P and P' lie on $l(m, d)$. Any other line through P and P' must be a line $l(m', d')$. Hence $m'a+d' = b$, $m'a'+d' = b'$, and $m'(a-a') = b-b'$. This yields $m = m'$ and $d' = b-m'a = b-ma = d$. Thus $l(m, d) = l(m', d')$.

In either case we have one and only one line connecting P and P'.

To Axiom A2 Let $P = (a, b)$ be any point and l a line. If $l = l(c)$, the line $l(a)$ is incident with P. By (α) it is parallel to l. By (β), $l(a)$ is the only line through P parallel to l.

If $l = l(m, d)$, the line $l' = l(m, b-ma)$ through P is parallel to l by (γ). Conversely, any line through P parallel to l is a line $l(m', d')$ by (β). By (γ), $l(m', d') \parallel l(m, d)$ implies $m' = m$. Since $P \in l(m', d')$, we obtain

$$d' = b-m'a = b-ma = d \quad \text{and} \quad l(m', d') = l(m, d).$$

To verify Axiom A3, choose the points $(0, 0)$, $(0, 1)$, and $(1, 0)$. They are readily shown to be non-collinear.

1.4
FINITE PLANES

In the examples (iv)–(vi), we studied triplets of *finite* point and line sets with an incidence relation. But we did not construct any *affine planes* with a finite number of points and lines. Such a plane is obtained if we try to construct an affine plane $\mathfrak{A} = (\mathbf{P}, \mathbf{L}, \mathbf{I})$ with a minimum number of points. By Axiom A3, the set $\mathbf{P}$ must contain a triangle ABC. Thus $\mathbf{L}$ contains its sides (6); cf. Axiom A1.

Applying Axiom A2, construct the lines d through C parallel to c and e through B parallel to b. Thus

$$d \parallel c \not\parallel b \parallel e \qquad \text{(i.e. } d \parallel c, c \not\parallel b, b \parallel e\text{).}$$

Hence $d \not\parallel e$ and these lines intersect at a point D.

The point D is distinct from A, B, C: since $C \not{I} c$, we have $d \neq c$. Hence $d \parallel c$ implies $D \neq A, B$, and similarly $D \neq C$.

Let $f = [A, D]$. The six lines a, b, c, d, e, f are mutually distinct. This is obvious for a, b, c. Furthermore, $d \neq a, b, c$. Otherwise either $a = d \parallel c$ or $b = d \parallel c$ (contradicting Theorem 6) or $C\,\mathbf{I}\,c = d$. Symmetrically, $e \neq a, b, c$. Since $c \not\parallel b \parallel e$, ABD is a triangle. Hence $B \not{I} f$. As B is incident with a, c, e, these lines must be distinct from f. Similarly, $f \neq b, d$.

Put

$$\mathbf{P} = \{A, B, C, D\}, \qquad \mathbf{L} = \{a, b, c, d, e, f\}.$$

Thus $\mathbf{P}$ and $\mathbf{L}$ are sets of four mutually distinct points and of six mutually distinct lines, respectively. The incidence relation is the following set of pairs

$$\mathbf{I} = \{(A, b),\ (A, c),\ (A, f),\ (B, a),\ (B, c),\ (B, e),$$
$$(C, a),\ (C, b),\ (C, d)\ (D, d)\ (D, e),\ (D, f)\}.$$

It is indicated in Figure 1.3 and in the following *incidence table*, the asterisks

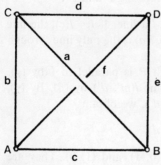

Figure 1.3

	a	b	c	d	e	f
A		*	*			*
B	*		*		*	
C	*	*		*		
D				*	*	*

1.4 FINITE PLANES

denoting incidence. It shows that Axiom A1 is satisfied. Also we have the following pairs of parallel lines

$a \parallel f, \quad b \parallel e, \quad c \parallel d.$

This implies A2. Finally A, B, C are not collinear. Thus

$\mathfrak{A}_2 = (\mathbf{P}, \mathbf{L}, \mathbf{I})$

is an affine plane.

In $\mathfrak{A}_2$, every line is incident with exactly two points. We call $\mathfrak{A}_2$ the *affine plane of order two*. Note that every point is incident with exactly three lines.

We wish to generalize our last remarks and to study the simplest properties of *finite planes*, i.e. of planes with only a finite number of points. We prepare our discussion by

LEMMA 9.1 *If one line is incident with at least n points, then every line is.*

PROOF Suppose the line a is incident with the n mutually distinct points $A_1, \ldots, A_n$. Let $b \neq a$. By Theorem 7(iii) and Theorem 2 there is a line l_1 through A_1 which is parallel to neither a nor b. Let l_k denote the line through A_k parallel to l_1. Thus the lines $l_1, \ldots, l_n$ are parallel. We have $l_k \nparallel a$ and $A_k = [l_k, a]$. Since

$[l_k, a] = A_k \neq A_j = [l_j, a] \qquad \text{for } k \neq j,$

Theorem 5 implies $l_k \neq l_j$. Thus the n parallel lines $l_1, \ldots, l_n$ are mutually distinct.

Furthermore $l_k \parallel l_1 \nparallel b$. Hence $l_k \nparallel b$ and Theorem 5 implies that the n points

$B_k = [l_k, b] \qquad (k = 1, \ldots, n)$

on b are mutually distinct. $\square$

THEOREM 9 *If the affine plane $\mathfrak{A}$ contains one line which is incident with a finite number of points only, then there exists an integer $n \geq 2$ such that every line is incident with exactly n points.*

PROOF Assume the line l is incident with exactly n points. Let h denote a second line. By our lemma, h is incident with at least n points. Suppose h is incident with not less than m points. Then our lemma implies that l also is incident with at least m points. Thus h is incident with exactly n points. $\square$

A plane has the *order n* if every line is incident with exactly n points.

THEOREM 10 *Every point in an affine plane of order n is incident with exactly $n+1$ lines.*

PROOF Given a point P in our plane, there exists, by Axiom A3, a line l not incident with P. Let $A_1, \ldots, A_n$ denote the points on l. Then $[P, A_1], \ldots, [P, A_n]$ and the line through P parallel to l are $n+1$ mutually distinct lines.

Conversely, any line through P intersects l at some point A_k unless it is parallel to l. Thus there are no other lines through P.

THEOREM 11 *Every parallel pencil in a plane of order n contains exactly n lines.*

PROOF Let Π_l denote any parallel pencil. By Theorem 6, there exists a line $h \nparallel l$. Through each point P of h there is exactly one line l_P of Π_l.

Conversely, each line of Π_l intersects h. Hence it is equal to one of the lines l_P where $P \, \mathrm{I} \, h$. Thus these are all the lines of Π_l.

If P and Q are distinct points on h, Theorem 5 yields $l_P \neq l_Q$. Hence the number of lines in Π_l is equal to the number n of points on h. $\square$

THEOREM 12 *An affine plane of order n contains exactly $n+1$ parallel pencils.*

PROOF Choose any point P. Let $a_1, \ldots, a_{n+1}$ denote the lines through P. The $n+1$ parallel pencils

$$\Pi_{a_1}, \ldots, \Pi_{a_{n+1}} \tag{10}$$

are mutually distinct by Theorem 4. Hence there are at least $n+1$ parallel pencils.

Conversely, any parallel pencil Π contains a line a through P by Axiom A2. Hence $\Pi = \Pi_a$ is equal to one of the pencils (10). $\square$

THEOREM 13 *An affine plane of order n contains exactly n^2 points and $n(n+1)$ lines.*

PROOF
(i) Let

$$\Pi_l = \{l_1, \ldots, l_n\}$$

denote a parallel pencil. By Axiom A2, each point is incident with exactly one line parallel to l, i.e. with precisely one of the n lines $l_1, \ldots, l_n$. By Theorem 9, each of them is incident with exactly n points. Thus the plane contains exactly $n \times n$ points.
(ii) By Theorem 12 there are precisely $n+1$ parallel pencils. By Theorem 10 each of them contains exactly n lines. As every line belongs to one and only one parallel pencil [cf. Theorem 4], we have altogether exactly $n(n+1)$ lines. $\square$

Note that by Theorem 9 an affine plane is finite if and only if it has order n for some n.

The question whether there are affine planes of a given order n is only partially solved. We know that there are such planes if n is the power of a prime number, and we do not know any planes whose order is not a power of a prime number. The famous Bruck-Ryser Theorem states: Let $n \equiv 1$ or $2 \pmod 4$ and suppose n is neither a prime power nor the sum of two squares. Then there is no affine plane of order n.

We thus know: (i) There are planes of the orders 2, 3, 4, 5, 7, 8, 9, 11, 13, 16, ... (ii) There are no planes of the orders 6, 14, 21, 22, ... But we do not know whether or not there are planes of the orders 10, 12, 15, ...

EXERCISES

1 Suppose the triplet **P**, **L**, $\mathbf{I}$, satisfies the following axiom and no other: Two distinct lines have not more than one point in common. (i) Show that two distinct points have not more than one connecting line. (ii) Show by an example that two distinct points need not have a connecting line.

2 Which sections of Theorems 2 and 3 would remain valid if we call two lines parallel if and only if they have no point in common.

3 Suppose the set **P** consists of the four vertices A, B, C, D of a regular tetrahedron in euclidean three-space, and **L** consists of its six edges

$$a = [B, C], \quad b = [C, A], \quad c = [A, B], \quad d = [C, D], \quad e = [B, D], \quad f = [A, D].$$

'Incidence' means ordinary incidence. Show that $\mathfrak{A} = (\mathbf{P}, \mathbf{L}, \mathbf{I})$ is an affine plane of order two [interpret Figure 1.3 as the drawing of a tetrahedron!].

4 Using Exercise 3, construct a triplet $(\mathbf{P}, \mathbf{L}, \mathbf{I})$ satisfying the axioms A1 and A3 but with a point A and a line l such that more than one line through A is parallel to l.

5 Let O be a point in euclidean three-space. The 'points' of **P** are the lines through O. The 'lines' of **L** are the planes through O. A 'point' is 'incident' with a 'line' if it is contained in the latter. Verify that the triplet $(\mathbf{P}, \mathbf{L}, \mathbf{I})$ satisfies the axioms A1 and A3 but that any two 'lines' always 'intersect' in exactly one 'point.' Check that these results are not affected by interchanging the definitions of 'points' and 'lines' if the definition of 'incidence' is suitably modified.

6 (i) Prove Theorem 6. (ii) Show that there is no point incident with all the sides of a triangle.

7 Suppose the affine plane $\mathfrak{A}$ contains a line with *at least n* points. Show that (i) every parallel pencil contains *at least n* lines; (ii) each point is incident with *at least* $n+1$ lines.

8 Given a finite affine plane of odd order n, show that any set of $n+2$ points contains three (mutually distinct) collinear ones.

9 Repeat the construction of a co-ordinate plane replacing $\mathbb{R}$ by any skew-field; cf. p. 108.

2
Collineations

2.1
BIJECTIONS

A *bijection* α of a set **S** onto a set **T** maps each element A of **S** onto an element

$$B = A\alpha \tag{1}$$

of **T** such that (i) $A\alpha = A'\alpha$ implies $A = A'$ if A, A' lie in **S**, (ii) to any $B \in \mathbf{T}$ there is an $A \in \mathbf{S}$ which solves (1).

By (i), A is uniquely determined by B. We call A the *inverse image* of B. By associating with each element B of **T** its inverse image A, we obtain a well-defined mapping α^{-1} from **T** into **S**, the *inverse* of α. Thus

$$B\alpha^{-1} = A \Leftrightarrow A\alpha = B \quad \text{for all } B \in \mathbf{T}.$$

Obviously, α^{-1} also is a bijection. It satisfies

$$(\alpha^{-1})^{-1} = \alpha. \tag{2}$$

A trivial but important example of a bijection is the *identical map* $\iota_\mathbf{S}$ from **S** onto itself defined through

$$A\iota_\mathbf{S} = A \quad \text{for all } A \in \mathbf{S}.$$

If α and β are bijections of **S** onto **T** and of **T** onto the set **U**, respectively, the mapping $\alpha\beta$ of **S** onto **U** is defined through

$$A\alpha\beta = (A\alpha)\beta \quad \text{for all } A \in \mathbf{S}.$$

Then $\alpha\beta$ is readily seen to be a bijection of **S** onto **U** with the inverse

$$(\alpha\beta)^{-1} = \beta^{-1}\alpha^{-1}. \tag{3}$$

We call $\alpha\beta$ the *product* of α by β. Note that

$$\iota_\mathbf{S}\alpha = \alpha\iota_\mathbf{T} = \alpha \tag{4}$$

and

$$\alpha\alpha^{-1} = \iota_\mathbf{S} \quad \text{and} \quad \alpha^{-1}\alpha = \iota_\mathbf{T}. \tag{5}$$

2.2 COLLINEATIONS

Our composition is associative. If γ is a bijection of $\mathbf{U}$ onto a set $\mathbf{V}$, then

$$(\alpha\beta)\gamma = \alpha(\beta\gamma) \quad \text{(cf. Exercise 1).} \tag{6}$$

2.2 COLLINEATIONS

Let $\mathfrak{A} = (\mathbf{P}, \mathbf{L}, \mathrm{I})$ and $\mathfrak{A}' = (\mathbf{P}', \mathbf{L}', \mathrm{I}')$ denote two affine planes. We call a *pair* α of bijections α_1 of $\mathbf{P}$ onto $\mathbf{P}'$ and α_2 of $\mathbf{L}$ onto $\mathbf{L}'$ a *collineation* of $\mathfrak{A}$ onto $\mathfrak{A}'$ if

$$P \mathrm{I} \, l \Leftrightarrow P\alpha_1 \, \mathrm{I}' \, l\alpha_2.$$

It is customary to drop the indices and to use the same letter α not only for the pair but also for the individual bijections. Furthermore, as a rule, the same symbol I is used to denote incidence in both $\mathfrak{A}$ and $\mathfrak{A}'$. Thus we usually write the preceding formula in the form

$$\boxed{P \mathrm{I} \, l \Leftrightarrow P\alpha \, \mathrm{I} \, l\alpha.} \tag{7}$$

This notation will be justified by Theorem 3.

The identical bijections $\iota_\mathbf{P}$ and $\iota_\mathbf{L}$ induce a collineation of $\mathfrak{A}$ onto itself, the *identity* $\iota = \iota_\mathfrak{A}$.

Another obvious but important example is the following: Given any affine plane $\mathfrak{A} = (\mathbf{P}, \mathbf{L}, \mathrm{I})$, define

$$l' = \{P \mid P \mathrm{I} \, l\} \quad \text{for each } l \in \mathbf{L}.$$

Thus $P \in l'$ if and only if $P \mathrm{I} \, l$. Put

$$\mathbf{L}' = \{l' \mid l \in \mathbf{L}\}$$

and define

$$\mathfrak{A}' = (\mathbf{P}, \mathbf{L}', \in).$$

Thus our incidence relation in $\mathfrak{A}'$ consists of all the pairs (P, l') such that $P \in l'$. We readily verify that $\mathfrak{A}'$ is an affine plane and that

$$P\alpha = P, \quad l\alpha = l' \quad \text{for all } P \in \mathbf{P} \text{ and all } l \in \mathbf{L}$$

defines a collineation α of $\mathfrak{A}$ onto $\mathfrak{A}'$.

Two affine planes $\mathfrak{A}$ and $\mathfrak{A}'$ are called *isomorphic* if there is a collineation of $\mathfrak{A}$ onto $\mathfrak{A}'$.

THEOREM 1 *Isomorphism of affine planes is an equivalence relation.*

PROOF
(i) *Reflexivity* The identity is a collineation of $\mathfrak{A}$ onto itself.
(ii) *Symmetry* Suppose the affine plane $\mathfrak{A}$ is isomorphic to the affine plane $\mathfrak{A}'$.

Then there is a collineation α of $\mathfrak{A}$ onto $\mathfrak{A}'$. It consists of two bijections, α_1 and α_2. By (7),

$$\alpha^{-1} = (\alpha_1^{-1}, \alpha_2^{-1}) \tag{8}$$

is a collineation of $\mathfrak{A}'$ onto $\mathfrak{A}$. Thus $\mathfrak{A}'$ is isomorphic to $\mathfrak{A}$.

(iii) *Transitivity* Assume there are collineations α of $\mathfrak{A}$ onto $\mathfrak{A}'$ and β of $\mathfrak{A}'$ onto $\mathfrak{A}''$. Let $\alpha = (\alpha_1, \alpha_2)$ and $\beta = (\beta_1, \beta_2)$. Then

$$\alpha\beta = (\alpha_1\beta_1, \alpha_2\beta_2) \tag{9}$$

is a pair of bijections. If P is a point and l a line in $\mathfrak{A}$, then $P \, \mathrm{I} \, l$ is by (7) equivalent to

$P\alpha \, \mathrm{I} \, l\alpha$ and $P\alpha\beta \, \mathrm{I} \, l\alpha\beta$.

Thus $\alpha\beta$ is a collineation from $\mathfrak{A}$ to $\mathfrak{A}''$. □

We call the collineation α^{-1} the *inverse* of α and $\alpha\beta$ the *product* of the collineations α and β; cf. (8) and (9).

THEOREM 2 *Let α be a collineation of an affine plane $\mathfrak{A}$. Let P, Q be points and l, h be lines of $\mathfrak{A}$. Then*
(i) $P \neq Q \Leftrightarrow P\alpha \neq Q\alpha$.
(ii) *If $P \neq Q$ then* $[P, Q]\alpha = [P\alpha, Q\alpha]$.
(iii) $l \parallel h \Leftrightarrow l\alpha \parallel h\alpha$ *and* $l \nparallel h \Leftrightarrow l\alpha \nparallel h\alpha$.
(iv) *If $l \nparallel h$ then* $[l, h]\alpha = [l\alpha, h\alpha]$.

PROOF
(i) is trivial.
(ii) Let $l = [P, Q]$. Then $l \, \mathrm{I} \, P, Q$. Hence by (7), $l\alpha \, \mathrm{I} \, P\alpha, Q\alpha$. Thus by (i), $l\alpha = [P\alpha, Q\alpha]$.
(iii) It is sufficient to prove the second statement. Let $l \nparallel h$. Thus there is exactly one point $P \, \mathrm{I} \, l, h$. By (7), $P\alpha \, \mathrm{I} \, l\alpha, h\alpha$. If $P' \, \mathrm{I} \, l\alpha, h\alpha$, then (7) yields $P'\alpha^{-1} \, \mathrm{I} \, l, h$ and hence $P'\alpha^{-1} = P$ or $P' = P\alpha$. Thus $l\alpha \nparallel h\alpha$.

Conversely, let $l\alpha \nparallel h\alpha$. Then from the above

$l = (l\alpha)\alpha^{-1} \nparallel (h\alpha)\alpha^{-1} = h$.

(iv) follows from (iii). □

Theorem 2(iii) implies

COROLLARY 2.1 *A collineation α of an affine plane maps a parallel pencil onto a parallel pencil; more precisely*

$\Pi_l \alpha = \Pi_{l\alpha}$.

Since α maps non-parallel lines onto non-parallel lines, we obtain from Corollary 2.1:

2.2 COLLINEATIONS

COROLLARY 2.2 *A collineation of the affine plane $\mathfrak{A}$ onto the affine plane $\mathfrak{A}'$ maps the set of the parallel pencils of $\mathfrak{A}$ bijectively onto that of $\mathfrak{A}'$.*

Let $\alpha = (\alpha_1, \alpha_2)$ denote a collineation of $\mathfrak{A} = (\mathbf{P}, \mathbf{L}, \mathbf{I})$ onto $\mathfrak{A}' = (\mathbf{P}', \mathbf{L}', \mathbf{I}')$. Suppose the points $A, B, C \in \mathbf{P}$ are collinear. Thus there is a line l incident with all three of them. By (7) this implies $l\alpha_2 \mathbf{I}' A\alpha_1, B\alpha_1, C\alpha_1$; i.e., the three image points are collinear. We say that collineations preserve collinearity. We wish to prove a converse of this remark.

THEOREM 3 *Given two affine planes $\mathfrak{A} = (\mathbf{P}, \mathbf{L}, \mathbf{I})$ and $\mathfrak{A}' = (\mathbf{P}', \mathbf{L}', \mathbf{I}')$ let α_1 be a bijection of $\mathbf{P}$ onto $\mathbf{P}'$ which preserves collinearity. Then there exists one and only one bijection α_2 of $\mathbf{L}$ onto $\mathbf{L}'$ such that $\alpha = (\alpha_1, \alpha_2)$ is a collineation.*

PROOF We first verify the uniqueness of α_2. Let β_2 also be a bijection of $\mathbf{L}$ onto $\mathbf{L}'$ such that $\beta = (\alpha_1, \beta_2)$ is a collineation. Put

$$\gamma = \alpha\beta^{-1} = (\alpha_1\alpha_1^{-1}, \alpha_2\beta_2^{-1}) \quad (\text{cf. (8) and (9)}).$$

Thus $\gamma = (\gamma_1, \gamma_2)$ is a collineation of $\mathfrak{A}$ onto itself. Here $\gamma_1 = \alpha_1\alpha_1^{-1} = \iota_{\mathbf{P}}$ is the identity mapping of $\mathbf{P}$ onto itself and $\gamma_2 = \alpha_2\beta_2^{-1}$. If $l \in \mathbf{L}$, choose any two distinct points P, Q incident with l. Thus $l = [P, Q]$. By Theorem 2 (ii),

$$l\gamma_2 = [P\gamma_1, Q\gamma_1] = [P, Q] = l.$$

Hence γ_2 maps every line of l onto itself, i.e. $\gamma_2 = \iota_{\mathbf{L}}$. This yields $\beta_2 = \gamma_2\beta_2 = \alpha_2\beta_2^{-1}\beta_2 = \alpha_2$.

It remains to construct a bijection α_2 of $\mathbf{L}$ onto $\mathbf{L}'$ such that $\alpha = (\alpha_1, \alpha_2)$ is a collineation.

Let $l = [A, B] = [C, D]$. Without loss of generality assume $C \neq A, B$. Thus A, B, C are mutually distinct collinear points and so are $A\alpha_1, B\alpha_1, C\alpha_1$. With A, C, D, the points $A\alpha_1, C\alpha_1, D\alpha_1$ are collinear. Hence

$$[A\alpha_1, B\alpha_1] = [A\alpha_1, C\alpha_1] = [C\alpha_1, D\alpha_1].$$

We may therefore define a map α_2 of $\mathbf{L}$ into $\mathbf{L}'$ through

$$[A, B]\alpha_2 = [A\alpha_1, B\alpha_1]. \tag{10}$$

It remains to prove that $\alpha = (\alpha_1, \alpha_2)$ is a collineation, i.e. that α satisfies (7) and that α_2 is bijective. We first verify part of (7), viz. that α preserves incidence: let $P \mathbf{I} l$, choose $Q \mathbf{I} l$, $Q \neq P$. Then $l = [P, Q]$ and by (10),

$$l\alpha_2 = [P, Q]\alpha_2 = [P\alpha_1, Q\alpha_1];$$

in particular $P\alpha_1 \mathbf{I}' l\alpha_2$.

Next α_2 is surjective, i.e. every line of $\mathbf{L}'$ is an image: let $l' = [A', B'] \in \mathbf{L}'$; thus $A' \neq B'$. Since α_1 is bijective, the points $A'\alpha_1^{-1}$ and $B'\alpha_1^{-1}$ of $\mathbf{P}$ are distinct. By (10) and (5), we have

17

COLLINEATIONS

$$[A'\alpha_1^{-1}, B'\alpha_1^{-1}]\alpha_2 = [A'\alpha_1^{-1}\alpha_1, B'\alpha_1^{-1}\alpha_1] = [A', B'] = l'.$$

Thus l' is the image of the line $[A'\alpha_1^{-1}, B'\alpha_1^{-1}]$.

In the next steps we show that α_2 is *injective*, i.e. that different lines of L have distinct images. Since α preserves incidence, $l \not\parallel h$ implies that $l\alpha_2$ and $h\alpha_2$ have points in common. Thus if $l\alpha_2$ and $h\alpha_2$ are distinct parallel lines, l and h must be parallel.

Assume first that the lines l_1 and l_2 of L are not parallel. Choose $h' \parallel l_1\alpha_2$, $h' \neq l_2\alpha_2$. Since α_2 is surjective, there is a line $h \in L$ such that $h' = h\alpha_2$. From the above, we have $h \parallel l_1$, and hence $h \not\parallel l_2$. Therefore h' and $l_2\alpha_2$ have points in common. As h' and $l_1\alpha_2$ have no points in common, this implies $l_1\alpha_2 \neq l_2\alpha_2$. Consider now two parallel lines l_1 and l_2. Choose $h \not\parallel l_1$. Then the points $P_k = [h, l_k]$ exist and are distinct; thus $h = [P_1, P_2]$. Since $l_k \not\parallel h$, we have

$$l_k\alpha_2 \neq h\alpha_2 = [P_1\alpha_1, P_2\alpha_1], \quad k = 1, 2.$$

The point $P_1\alpha_1$ being incident with $l_1\alpha_2$, we obtain

$$P_2\alpha_1 \not\mathrel{I} l_1\alpha_2.$$

As $P_2\alpha_1 \mathrel{I} l_2\alpha_2$, this yields $l_1\alpha_2 \neq l_2\alpha_2$.

We have now proved that L have distinct images, and the proof that α_2 is bijective is complete. We still have to verify the second half of (7), viz. that $P\alpha_1 \mathrel{I} l\alpha_2$ implies $P \mathrel{I} l$.

Suppose $P \not\mathrel{I} l$. Choose a point $Q \mathrel{I} l$; thus $P \neq Q$. Put $h = [P, Q]$. Then $h \neq l$ and

$$h\alpha_2 = [P\alpha_1, Q\alpha_1] \neq l\alpha_2.$$

Since $P \neq Q = [h, l]$, we have $P\alpha_1 \neq Q\alpha_1 = [h\alpha_2, l\alpha_2]$. As $P\alpha_1 \mathrel{I} h\alpha_2$, this implies $P\alpha_1 \not\mathrel{I} l\alpha_2$. □

The last part of our proof yields

COROLLARY 3.1 *Given two affine planes* $\mathfrak{A} = (\mathbf{P}, \mathbf{L}, \mathrm{I})$ *and* $\mathfrak{A}' = (\mathbf{P}', \mathbf{L}', \mathrm{I}')$, *a pair of bijections of* P *onto* P' *and of* L *onto* L' *which preserves incidence is a collineation.*

From now on we study the collineations of an affine plane onto itself.

THEOREM 4. *The collineations* α *of an affine plane form a group with the unit element* ι.

PROOF The product of two collineations was defined in (9). The associativity of the composition of bijections induces that of the multiplication of collineations. By (4) and (9), we have $\iota\alpha = \alpha\iota = \alpha$. Finally by (8), $\alpha\alpha^{-1} = \alpha^{-1}\alpha = \iota$. □

2.3
FIXED ELEMENTS

We call the point P a *fixed point* of the collineation α if $P\alpha = P$. Similarly if $l\alpha = l$, l is called a *fixed line of* α. The *fixed set* $\mathbf{F}(\alpha)$ of α consists of all the fixed points and lines. Thus

$$\mathbf{F}(\iota) = \mathbf{P} \cup \mathbf{L}, \tag{11}$$

$$\mathbf{F}(\alpha^{-1}) = \mathbf{F}(\alpha), \tag{12}$$

$$\mathbf{F}(\alpha) \subset \mathbf{F}(\alpha^n) \quad \text{for } n = 2, 3, \ldots \tag{13}$$

We collect three basic properties of $\mathbf{F}(\alpha)$.

THEOREM 5
(i) *If P and Q are two distinct fixed points of α, then $[P, Q]$ is a fixed line of α.*
(ii) *If l and h are non-parallel fixed lines of α, then $[l, h]$ is a fixed point of α.*
(iii) *If P is a fixed point and l is a fixed line of α, then the line through P parallel to l is a fixed line of α.*

PROOF Suppose $P\alpha = P$, $Q\alpha = Q$, and $P \neq Q$. Then by Theorem 2(ii), $[P, Q]\alpha = [P\alpha, Q\alpha] = [P, Q]$.

Part (ii) follows similarly.

Finally, let h denote the line through P parallel to l. By Theorem 2, α preserves not only incidence but also parallelism. Hence $h\alpha$ is the line through $P\alpha = P$ parallel to $l\alpha = l$. Thus $h\alpha = h$. □

COROLLARY 5.1 *The triplet*

$$(\mathbf{F}(\alpha) \cap \mathbf{P}, \mathbf{F}(\alpha) \cap \mathbf{L}, \mathtt{I}') \tag{14}$$

satisfies the axioms A1 *and* A2. *Here* $\mathtt{I}'$ *is the restriction of* $\mathtt{I}$ *to* $(\mathbf{F}(\alpha) \cap \mathbf{P}) \times (\mathbf{F}(\alpha) \cap \mathbf{L})$. *Note that, e.g.,* $\mathbf{F}(\alpha) \cap \mathbf{P}$ *is the set of the fixed points of* α.

Axiom A1 for the triplet (14) follows from Theorem 5(i). Similarly, Axiom A2 for (14) is obtained from Axiom A2 for the whole plane and Theorem 5(iii). □

If $\mathbf{P}' \subset \mathbf{P}$, $\mathbf{L}' \subset \mathbf{L}$, and if $\mathfrak{A}' = (\mathbf{P}', \mathbf{L}', \mathtt{I}')$ satisfies the axioms A1–A3, $\mathfrak{A}'$ is called a *subplane* of $\mathfrak{A}$. Corollary 5.1 implies

COROLLARY 5.2 *If $\mathbf{F}(\alpha)$ contains three non-collinear points, (14) is a subplane of $\mathfrak{A}$.*

We next verify a simple condition for two collineations α and β to *commute*, i.e. for $\alpha\beta = \beta\alpha$.

THEOREM 6 *If the collineations α and β commute, α must map the fixed set of β onto itself.*

PROOF Let $\alpha\beta = \beta\alpha$. Suppose, e.g., that P is a fixed point of β. Then

$$P\alpha = (P\beta)\alpha = P(\beta\alpha) = P(\alpha\beta) = (P\alpha)\beta.$$

Thus $P\alpha \in \mathbf{F}(\beta)$ and α maps $\mathbf{F}(\beta)$ into itself.

Since $\alpha\beta = \beta\alpha$ implies

$$\alpha^{-1}\beta = \alpha^{-1}(\beta\alpha)\alpha^{-1} = \alpha^{-1}(\alpha\beta)\alpha^{-1} = \beta\alpha^{-1},$$

α^{-1} and β also commute and α^{-1} maps $\mathbf{F}(\beta)$ into itself. Hence $\mathbf{F}(\beta)$ is mapped by α onto itself. □

An *axis* of the collineation α is a line, all the points of which are fixed. If every line through a point is fixed, the point is called a *centre* of α. Thus every axis is a fixed line and every centre is a fixed point; cf. Theorem 5. If $\alpha = \iota$, every line is an axis and every point is a centre. We next show that a collineation $\neq \iota$ cannot have too many axes and centres.

THEOREM 7 *Let α be a collineation, $\alpha \neq \iota$.*
(i) *If α has a centre O, then it has no other fixed point.*
(ii) *If α has a centre O, then any fixed line is incident with O.*
(iii) *If α has an axis a, then any fixed point is incident with a.*

PROOF
(i) Suppose α also has the fixed point $F \neq O$. We first verify that F is a centre: let $l \mathbin{\mathrm{I}} F$. The line through O parallel to l being fixed, l itself is fixed by Theorem 5(iii). Thus every line through F is fixed.

Next let P be any point. If $P \mathbin{\not{\mathrm{I}}} [O, F]$, P is the intersection of the fixed lines $[O, P]$ and $[F, P]$ and hence fixed. From the above, P is even a centre.

Finally, if $P \mathbin{\mathrm{I}} [O, F]$ but $P \neq O, F$, choose any $Q \mathbin{\not{\mathrm{I}}} [O, F]$ and repeat the preceding construction replacing F by Q. Thus every point is a centre and every line is fixed. Hence $\alpha = \iota$.
(ii) Suppose there is a fixed line $f \mathbin{\not{\mathrm{I}}} O$. Let $P \mathbin{\mathrm{I}} f$. Then P is the intersection of the fixed lines $[O, P]$ and f. By Theorem 5(ii), P is a fixed point of α. This is excluded by the first part of our theorem.
(iii) Suppose there is a fixed point $O \mathbin{\not{\mathrm{I}}} a$ of α. By Theorem 5(iii), the line through O parallel to a is fixed. Let $l \mathbin{\mathrm{I}} O$, $l \nparallel a$. Since O and $[l, a]$ are fixed points, $l = [O, [l, a]]$ is fixed by Theorem 5(i). Thus every line through O is fixed and O is a centre. Since $a \mathbin{\not{\mathrm{I}}} O$, this is excluded by the second part of our theorem. □

COROLLARY 7.1 *The total number of centres and axes of a collineation $\neq \iota$ is not greater than one.*

PROOF Let $\alpha \neq \iota$. If α has a centre, Theorem 7(i) implies that α cannot have an axis or another centre. By Theorem 7(iii), α cannot have two axes.

REMARK We shall see later that a collineation $\neq \iota$ may have other fixed lines in addition to an axis without being the identity.

2.4
HOMOTHETIES

The collineation η is a *homothety* if

$$l\eta \parallel l \quad \text{for all } l \in \mathbf{L}. \tag{15}$$

If $\eta = (\eta_1, \eta_2)$ is a homothety, our definition implies

$$[P, Q] \parallel [P\eta_1, Q\eta_1] \quad \text{for every pair } P \neq Q. \tag{16}$$

We show that (16) characterizes the homotheties:

THEOREM 8 *Given an affine plane* $\mathfrak{A} = (\mathbf{P}, \mathbf{L}, \mathbf{I})$, *let* η_1 *denote a bijection of* $\mathbf{P}$ *onto itself which satisfies* (16). *Then there exists one and only one bijection* η_2 *of* $\mathbf{L}$ *onto itself such that* $\eta = (\eta_1, \eta_2)$ *is a collineation.* η *then is a homothety.*

PROOF If we can show that η_1 preserves collinearity, Theorem 3 implies the existence of a unique η_2 such that $\eta = (\eta_1, \eta_2)$ is a collineation. If $l = [P, Q]$ is any line, then

$$l\eta_2 = [P, Q]\eta_2 = [P\eta_1, Q\eta_1] \parallel [P, Q] = l.$$

Thus η will be a homothety.

Assume then that P, Q, R are collinear. We may also assume that they are mutually distinct. By (16)

$$[P\eta_1, Q\eta_1] \parallel [P, Q] = [P, R] \parallel [P\eta_1, R\eta_1].$$

Hence $[P\eta_1, Q\eta_1] = [P\eta_1, R\eta_1]$, and $P\eta_1$, $Q\eta_1$, $R\eta_1$ are collinear. Thus η_1 preserves collinearity. □

THEOREM 9 *The homotheties of the affine plane* $\mathfrak{A}$ *form a normal subgroup H of the group of the collineations of* $\mathfrak{A}$.

PROOF If η and η' satisfy (15), then $(l\eta)\eta' \parallel l\eta \parallel l$ for all $l \in \mathbf{L}$. Since the collineation η^{-1} preserves parallelism, $l\eta \parallel l$ implies $l = (l\eta)\eta^{-1} \parallel l\eta^{-1}$ for all $l \in \mathbf{L}$. Thus H is a subgroup.

Next, let α denote any collineation. Then by (15)

$$(l\alpha^{-1})\eta \parallel l\alpha^{-1}$$

and hence by Theorem 2(iii),

$$((l\alpha^{-1})\eta)\alpha \parallel (l\alpha^{-1})\alpha \quad \text{or} \quad l\alpha^{-1}\eta\alpha \parallel l$$

for all $l \in \mathbf{L}$. Thus $\alpha^{-1}\eta\alpha$ also is a homothety. □

THEOREM 10 *A fixed point of a homothety is a centre.*

PROOF Let $O\eta = O$; $l \mathbin{\text{I}} O$. Then $l\eta \mathbin{\text{I}} O\eta = O$ and by (15), $l\eta \parallel l$. Hence $l\eta = l$. $\square$

Theorem 10 and Corollary 7.1 imply

COROLLARY 10.1 *A homothety $\neq \iota$ has not more than one fixed point. In particular, it has no axis.*

If α is any collineation and if $P \neq P\alpha$, the line $[P, P\alpha]$ is called a *trace* of α. Every fixed line is either an axis or a trace. For if the fixed line f is incident with a point $P \neq P\alpha$, then $P\alpha \mathbin{\text{I}} f\alpha = f$ and therefore $f = [P, P\alpha]$.

Let η again denote a homothety. By Corollary 10.1 every fixed line of η is a trace. Conversely, let $l = [P, P\eta]$ be a trace of η. Then $l \mathbin{\text{I}} P$ implies $l\eta \mathbin{\text{I}} P\eta$. Thus by (15), $l\eta$ is the line through $P\eta$ parallel to l. Since $l \mathbin{\text{I}} P\eta$, this implies $l\eta = l$. We thus have

THEOREM 11 *A line is the trace of a homothety $\neq \iota$ if and only if it is fixed.*

We cannot deduce the existence of homotheties other than the identity from our axioms. On the other hand, it is easy to show that there cannot be too many homotheties.

THEOREM 12 *Let $A \neq B$. Then there is not more than one homothety mapping A onto C and B onto D.*

PROOF Suppose the homotheties η and η' satisfy

$$A\eta = A\eta' = C, \qquad B\eta = B\eta' = D.$$

By Theorem 7, $\eta'\eta^{-1}$ is a homothety. It has the fixed points A and B. Hence by Corollary 10.1, $\eta'\eta^{-1} = \iota$ and $\eta' = \eta$. $\square$

Concluding this section we study the set $H(a)$ of all the homotheties with the given fixed line a.

THEOREM 13 *For any line a, $H(a)$ is a subgroup of H. Let α be any collineation. Then*

$$\eta \mapsto \eta^\alpha (= \alpha^{-1}\eta\alpha), \qquad \eta \in H(a), \tag{17}$$

defines an isomorphism of $H(a)$ onto $H(a\alpha)$.

PROOF Let η', $\eta \in H(a)$. Since a is a fixed line of η' and of η^{-1}, it is one of $\eta'\eta^{-1}$. Thus $\eta'\eta^{-1} \in H(a)$.

Next consider the map (17). Since

$$\eta' = \alpha^{-1}\eta\alpha \Leftrightarrow \eta = \alpha\eta'^{-1}\alpha^{-1},$$

it is a bijection of $H(a)$ onto $H(a\alpha)$. Also

$$(\eta\eta')^\alpha = \alpha^{-1}\eta\eta'\alpha = \alpha^{-1}\eta\alpha \cdot \alpha^{-1}\eta'\alpha = \eta^\alpha \eta'^\alpha.$$

Thus (17) defines an isomorphism of $H(a)$ onto $H(a\alpha)$. □

THEOREM 14 *Given a line a and two points A and B not on a, there is not more than one homothety in $H(a)$ which maps A onto B.*

PROOF Let $\eta, \eta' \in H(a)$; $A\eta = A\eta' = B$. By Theorem 13, $\eta'\eta^{-1} \in H(a)$. Since $A\eta'\eta^{-1} = A$, Theorem 10 implies that A is a centre of the homothety $\eta'\eta^{-1}$. Since a is a fixed line of $\eta'\eta^{-1}$ and $A \not\in a$, Theorem 7(ii) implies $\eta'\eta^{-1} = \iota$. Thus $\eta' = \eta$. □

2.5
TRANSLATIONS

By Corollary 10.1, there are two types of homotheties $\neq \iota$, those with one fixed point and those with none. We first study the latter, defining: a *translation* is a homothety which either has no fixed points or is the identity. If the translation τ is not the identity, the fixed set $\mathbf{F}(\tau)$ consists of the traces of τ; cf. Theorem 11.

THEOREM 15 *If the translation τ is not the identity, $\mathbf{F}(\tau)$ is a parallel pencil Π.*

We call τ a translation *in the direction of* Π.

PROOF Through each point P there is a trace $[P, P\tau]$, i.e. a fixed line; cf. Theorem 11. Since τ has no fixed points, Theorem 5(ii) implies that any two fixed lines are parallel. Hence the set of fixed lines is a parallel pencil. □

We next verify a uniqueness theorem for translations.

THEOREM 16 *For every pair of points A, B there exists at most one translation τ which maps A onto B.*

PROOF Let τ and τ' denote two translations; $A\tau = A\tau' = B$. If $B = A$, we have $\tau = \tau' = \iota$. Let $B \neq A$. Then $a = [A, B]$ is a trace and hence a fixed line of τ and τ'; cf. Theorem 11. Thus τ and τ' belong to $H(a)$ and our assertion follows by applying Theorem 14 to A, B, and a line $a' \parallel a$, $a' \neq a$. □

Alternatively, we can prove Theorem 16 by a construction which will become important later.

Let τ be a translation, $A \neq B = A\tau$. If $C \not\in [A, B]$, then $[C, C\tau] \parallel [A, B]$ by Theorem 15. Furthermore, by Theorem 2(ii) and (15),

$$[B, C\tau] = [A\tau, C\tau] = [A, C]\tau \parallel [A, C].$$

Since $[A, C] \not\parallel [A, B]$, this determines the point $C\tau$: it is the intersection of the lines through C parallel to $[A, B]$ and through B parallel to $[A, C]$.

If $C \mathrm{I} [A, B]$, we repeat our construction, replacing A by some point D

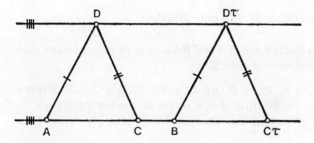

Figure 2.1

outside $[A, B]$ and B by $D\tau$. Thus the image of each point is uniquely determined. The same will therefore apply to the lines; cf. Figure 2.1.

THEOREM 17 *The translations form a normal subgroup T of the group of all the collineations.*

PROOF Let τ and τ' be translations; α is any collineation. We have to verify that τ^{-1}, $\tau'\tau$, and $\alpha^{-1}\tau\alpha$ are translations. By Theorem 9, they are homotheties. As $\mathbf{F}(\tau^{-1}) = \mathbf{F}(\tau)$, τ^{-1} has no fixed points unless $\tau = \iota$.

Suppose $P\tau'\tau = P$. Then $P\tau' = P\tau^{-1}$. Since τ^{-1} also is a translation, Theorem 16 implies $\tau' = \tau^{-1}$ or $\tau'\tau = \iota$. Thus either $\tau'\tau$ has no fixed point or $\tau'\tau = \iota$.

Finally, $P\alpha^{-1}\tau\alpha = P$ implies $(P\alpha^{-1})\tau = P\alpha^{-1}$. Thus τ has the fixed point $P\alpha^{-1}$ and $\tau = \iota$. Hence $\alpha^{-1}\tau\alpha = \iota$. □

Let $T(\Pi)$ denote the set of the translations in the direction of Π. Thus

$$T(\Pi) = \{\tau \in T \mid \mathbf{F}(\tau) = \Pi \text{ or } \tau = \iota\}.$$

Our definition implies

$$T(\Pi) \cap T(\Pi') = \{\iota\} \quad \text{if } \Pi \neq \Pi'. \tag{18}$$

THEOREM 18 *For any parallel pencil Π, $T(\Pi)$ is a normal subgroup of H, in particular of T. For every collineation α,*

$$\tau \mapsto \tau^\alpha = \alpha^{-1}\tau\alpha \quad \text{for all } \tau \in T(\Pi)$$

defines an isomorphism of $T(\Pi)$ onto $T(\Pi\alpha)$.

PROOF If both τ and τ' belong to $T(\Pi)$, then by Theorem 17, $\tau'\tau \in T$. Furthermore,

$$l\tau'\tau = l\tau = l \quad \text{for every } l \in \Pi.$$

Hence $\tau'\tau \in T(\Pi)$. Since $\mathbf{F}(\tau) = \mathbf{F}(\tau^{-1})$, $\tau \in T(\Pi)$ implies $\tau^{-1} \in T(\Pi)$. Thus $T(\Pi)$ is a subgroup of T.

2.5 TRANSLATIONS

Let α be any collineation. By Theorem 17 $\alpha^{-1}\tau\alpha$ is a translation. If $l \in \Pi$, then $(l\alpha)(\alpha^{-1}\tau\alpha) = l\tau\alpha = l\alpha$.
Hence

$$\alpha^{-1}\tau\alpha \in T(\Pi\alpha);$$

cf. Corollary 2.1.
    For any homothety α, $\Pi\alpha = \Pi$. Thus $T(\Pi)$ is a normal subgroup of H. The second statement is obvious; cf. the proof of Theorem 13. □

A fairly weak additional assumption yields some interesting results on the group T.

THEOREM 19 *If there are translations $\neq \iota$ in distinct directions, T is abelian.*

PROOF
(i) Let τ and τ' denote two translations $\neq \iota$ in distinct directions, say

$$\tau \in T(\Pi), \quad \tau' \in T(\Pi'), \quad \Pi \neq \Pi'. \tag{19}$$

By Theorem 18, τ'^{-1} and $\tau^{-1}\tau'\tau$ both lie in $T(\Pi')$. Hence

$$\tau'^{-1}(\tau^{-1}\tau'\tau) \in T(\Pi').$$

On the other hand, $T(\Pi)$ contains $\tau'^{-1}\tau^{-1}\tau'$ and τ and hence $(\tau'^{-1}\tau^{-1}\tau')\tau$. Thus

$$\tau'^{-1}\tau^{-1}\tau'\tau \in T(\Pi) \cap T(\Pi') = \{\iota\}$$

or

$$(\tau\tau')^{-1}(\tau'\tau) = \iota \quad \text{or} \quad \tau'\tau = \tau\tau';$$

cf. (18).
(ii) Let τ, τ' denote two translations in the same direction Π. By our assumption there is a $\tau'' \notin T(\Pi)$. Since $\tau'\tau'' \in T(\Pi)$ would imply $\tau'' = \tau'^{-1}(\tau'\tau'') \in T(\Pi)$, we have

$$\tau'\tau'' \notin T(\Pi).$$

Hence by (i),

$$\tau\tau' = \tau(\tau'\tau'')\tau''^{-1} = (\tau'\tau'')\tau\tau''^{-1} = \tau'\tau\tau'\tau''^{-1} = \tau'\tau. \square$$

THEOREM 20 *Suppose there are translations $\neq \iota$ in distinct directions. If there is one translation $\neq \iota$ of finite order, then every translation has finite order and there is a prime number p such that the order of every translation $\neq \iota$ is equal to p.*

PROOF Let $\tau_0 \neq \iota$ be a translation of finite order n and let p be a prime divisor of n; thus $n = pm$. Then the order of $\tau = \tau_0^m$ is equal to p.
    Let τ and τ' satisfy (19), $\tau' \neq \iota$. Thus

$$\Pi'' = \mathbf{F}(\tau\tau') \neq \Pi' \quad \text{and} \quad \Pi'' \subset \mathbf{F}((\tau\tau')^p).$$

By Theorem 19,

$$(\tau\tau')^p = \tau^p \tau'^p = \tau'^p.$$

Hence $F(\tau'^p)$ contains the distinct parallel pencils Π' and Π''. This yields $\tau'^p = \iota$. Thus every translation $\tau' \in T(\Pi)$ is of order p.

Repeating our argument starting from τ', we obtain that every translation not in $T(\Pi')$, in particular every element $\neq \iota$ of $T(\Pi)$ also has the order p. □

2.6
DILATATIONS

We call a homothety with a fixed point a *dilatation* δ. If $\delta \neq \iota$, we obtain from Theorem 10 and Corollary 10.1 that δ has exactly one fixed point O, and O is a centre of δ. By Theorem 7, δ has no fixed lines other than those through O. Thus

$$F(\delta) = \{O\} \cap \{l \mid l\mathrm{I}\, O\}.$$

By Theorem 11, the traces of δ are its fixed lines.

THEOREM 21 *For any point O, the set $D(O)$ of the dilatations with the centre O is a group. If α is any collineation, the mapping*

$$\delta \mapsto \delta^\alpha = \alpha^{-1}\delta\alpha, \qquad \delta \in D(O),$$

defines an isomorphism of $D(O)$ onto $D(O\alpha)$.

The proof is obvious. From Theorem 12 we obtain:

THEOREM 22 *If A and B are distinct from O, there is not more than one dilatation with the centre O which maps A onto B. If there is one, the points O, A, B must be collinear.*

By Theorem 9, the product $\eta\eta'$ of two homotheties η, η' is a homothety. A homothety was either a translation or a dilatation. We obtain special cases by prescribing for each of the homotheties η, η', $\eta\eta'$ whether it shall be a translation or a dilatation. However, not all of these cases are possible.

THEOREM 23
(i) *The product of two translations is a translation.*
(ii) *The product of a translation τ and a dilatation $\delta \neq \iota$ is a dilatation.*
(iii) *The product of two dilatations with the same centre is a dilatation.*
(iv) *If the product of the two dilatations δ, $\delta' \neq \iota$ with distinct centres O, O', respectively, is a translation, then $\delta\delta''$ is a dilatation for every $\delta'' \neq \delta'$ in $D(O')$.*

REMARK Suppose the groups $D(O)$ and $D(O')$ are linearly transitive, $O \neq O'$; cf. Chapter 3. Then there exists to every $\delta \in D(O)$ a dilatation $\delta' \in D(O')$ such that $\delta\delta'$ is a translation; cf. Chapter 4, Exercise 3.

2.6 DILATATIONS

PROOF OF THEOREM 23 The assertions (i) and (iii) have been proved in Theorems 17 and 21.

To (ii). Assume $\tau\delta$ is a translation. Then $\delta = \tau^{-1}(\tau\delta)$ would be the product of two translations and therefore a translation; contradiction.

To (iv). Put $\delta\delta' = \tau$. Assume $\delta\delta'' = \tau'$ is a translation for some δ'' in $D(O')$. Then

$$\delta'^{-1}\delta'' = \delta'^{-1}\delta^{-1}\delta\delta'' = \tau^{-1}\tau'$$

is a translation. Since it has the fixed point O', it must be the identity. Hence $\delta' = \delta''$. □

A collineation α is an *involution* [is *involutory*] if

$$\alpha^2 = \iota, \qquad \alpha \neq \iota.$$

A *reflection* σ in the point O is an involutory dilatation with the centre O; thus

$$\sigma \in D(O), \qquad \sigma^2 = \iota, \qquad \sigma \neq \iota.$$

THEOREM 24
(i) *The product of a translation with a reflection in a point is a reflection in a point.*
(ii) *The product of two reflections in distinct points is a translation.*
(iii) *The product of three reflections in points is again a reflection in a point, unless all the three points coincide.*
(iv) *There is at most one reflection in the point O if there is a collineation α such that $O\alpha \neq O$.*
(v) *If there are involuntary translations, there are no reflections in points.*

PROOF *To* (i). Let τ be a translation and let σ denote a reflection in a point. With τ, $\sigma^{-1}\tau\sigma$ is a translation; cf. Theorem 17. Hence

$$(\tau\sigma)^2 = \tau(\sigma\tau\sigma) = \tau(\sigma^{-1}\tau\sigma)$$

is a translation. By Theorem 23(ii), $\tau\sigma$ is a dilatation. Thus by Theorem 21, $(\tau\sigma)^2$ also is a dilatation. This yields $(\tau\sigma)^2 = \iota$. If $\tau\sigma = \iota$, $\sigma = \tau^{-1}$ would be a translation $\neq \iota$. Thus $\tau\sigma \neq \iota$ and $\tau\sigma$ is an involution.

To (ii). Let σ, σ' be two reflections in distinct points. Thus $\sigma \neq \sigma'$ and $\sigma\sigma' = \sigma^{-1}\sigma' \neq \iota$. We have to show that the homothety $\sigma\sigma'$ is not a dilatation.

Suppose $\sigma\sigma'$ has the fixed point O. Then

$$O\sigma\sigma' = O \quad \text{or} \quad O\sigma = O\sigma' = Q, \text{ say.} \tag{20}$$

Then $Q\sigma' = O\sigma'^2 = O$ and

$$Q\sigma\sigma' = O\sigma^2\sigma' = O\sigma' = Q.$$

Thus $\sigma\sigma'$ would have the two fixed points O and Q. Since $\sigma\sigma' \neq \iota$, this implies $O = Q$; cf. Corollary 10.1. By (20), O would be a common fixed point of σ and σ'; contradiction.

To (iii). Let $\sigma, \sigma', \sigma''$ denote reflections in O, O', O'', respectively. Thus either $O \neq O'$ or $O' \neq O''$. In the first [second] case, $\sigma\sigma'$ [$\sigma'\sigma''$] is a translation by (ii) Hence by (i), $\sigma\sigma'\sigma''$ is a reflection in a point.

To (iv). Let σ and σ' denote two reflections in O. By Theorem 21, $\sigma'' = \alpha^{-1}\sigma\alpha$ is a reflection in $O\alpha \neq O$. By (ii), $\sigma\sigma''$ and $\sigma''\sigma'$ are translations and so is their product $\sigma\sigma' = \sigma\sigma'' \cdot \sigma''\sigma'$. As $\sigma\sigma'$ has the fixed point O, it is the identity. Hence $\sigma' = \sigma^{-1} = \sigma$.

To (v). Assume we had both an involuntary translation τ and a reflection σ in a point. By (i), $\tau\sigma$ would be a reflection in some point. Thus $\iota = (\tau\sigma)^2$ or

$$\tau\sigma = (\tau\sigma)^{-1} = \sigma^{-1}\tau^{-1} = \sigma\tau.$$

Hence $\sigma = \tau\sigma\tau = \tau^{-1}\sigma\tau$ would be both a reflection in some point O and in $O\tau \neq O$; contradiction. □

COROLLARY 24.1 *If there is a reflection σ in a point, every translation τ can be expressed as the product of σ by another reflection in a point.*

PROOF By Theorem 24(i), $\sigma' = \sigma\tau$ is a reflection in a point. We obtain $\tau = \sigma^{-1}\sigma' = \sigma\sigma'$. □

2.7
AXIAL AFFINITIES

We conclude this chapter by briefly discussing another class of collineations, viz. those with an axis. The results which apply to these *axial affinities* are mostly similar to those on homotheties. It will be by imbedding an affine plane into a projective one that we shall understand the relationship between these two types of collineations; cf. Chapter 7.

We formally define an *affinity with the axis a* as a collineation with the axis a.

THEOREM 25 [cf. Theorem 9] *The affinities with the axis a form a subgroup $A(a)$ of the group of the collineations. If γ is any collineation, then*

$$\alpha \mapsto \alpha^\gamma = \gamma^{-1}\alpha\gamma \qquad \alpha \in A(a)$$

defines an isomorphism of $A(a)$ onto $A(a\gamma)$.

The proof is obvious.

THEOREM 26 [cf. Theorem 11] *Let $\alpha \in A(a)$, $\alpha \neq \iota$. Then a line $\neq a$ is a trace of α if and only if it is a fixed line.*

PROOF Obviously, this condition is sufficient. Conversely, let $l = [P, P\alpha]$ be a

2.7 AXIAL AFFINITIES

trace. If $l \not\parallel a$, then $F = [l, a]$ is a fixed point. We have $[F, P] = l = [F, P\alpha]$ and therefore

$$l\alpha = [F, P]\alpha = [F\alpha, P\alpha] = [F, P\alpha] = l.$$

If $l \parallel a$, then $l\alpha \parallel a\alpha = a$. Thus $l\alpha$ is the line through $P\alpha$ parallel to a, i.e $l\alpha = l$; cf. Exercise 5. □

The preceding theorem enables us to determine the fixed set $\mathbf{F}(\alpha)$ of an axial affinity α.

THEOREM 27 [cf. Theorem 15] *Let $\alpha \in A(a)$, $\alpha \neq \iota$. Then $\mathbf{F}(\alpha)$ consists of the points of a, the line a, and a parallel pencil.*

PROOF Clearly, the line a and the points of a belong to $\mathbf{F}(\alpha)$. By Theorem 7, α has no fixed points outside a. It remains to determine the lines in $\mathbf{F}(\alpha)$.

By Theorem 26, the fixed lines of α are identical with the traces. We first assume that every fixed line of α is parallel to a. Then every line parallel to and distinct from a is a trace and therefore a fixed line. Thus $\mathbf{F}(a)$ consists of Π_a and the points of a.

Assume next that there is a fixed line $l \not\parallel a$. Then every line $h \in \Pi_l$ is the line parallel to l through the fixed point (h, a) and hence fixed. Thus $\Pi_l \subset \mathbf{F}(\alpha)$.

Conversely assume $h \in \mathbf{F}(\alpha)$. Then $h \parallel l$ or $h = a$. For otherwise α would have fixed points outside a. This proves our assertion. □

COROLLARY 27.1 *The set of the traces of α is a parallel pencil, possibly omitting the line a.*

THEOREM 28 [cf. Theorem 16] *Let B and C be non-incident with a. Then there is not more than one $\alpha \in A(a)$ such that $B\alpha = C$.*

PROOF See Exercise 16.

We finally discuss the commutativity of homotheties and axial affinities in a few cases. One case was dealt with in Theorem 19. A more systematic treatment will be found in Chapter 7.

THEOREM 29
(i) *The product of a translation $\tau \neq \iota$ and a dilatation $\delta \neq \iota$ is not commutative.*
(ii) *The product of two dilatations δ, $\delta' \neq \iota$ with distinct centres O and O', respectively, is not commutative.*
(iii) *The product of a homothety $\eta \neq \iota$ and an affinity $\alpha \neq \iota$ with the axis a is commutative if and only if a is a trace of η.*

PROOF (i) and (ii) are corollaries of Theorem 6.

To (iii). If $\alpha\eta = \eta\alpha$, η must map the set $\{P \mid P \mathbin{\mathrm{I}} a\}$ of the fixed points of α onto itself, by Theorem 6. Hence $\eta a = a$.

Conversely, let a be a trace of η. Then $P\eta \mathbin{\mathrm{I}} a$ and

$$P\alpha^{-1}\eta\alpha\eta^{-1} = P\eta\alpha\eta^{-1} = P\eta\eta^{-1} = P \quad \text{for all } P\,\mathrm{I}\,a. \tag{21}$$

By Theorem 8, $\alpha^{-1}\eta\alpha$ is a homothety. Thus $(\alpha^{-1}\eta\alpha)\eta^{-1}$ is one too. By (21) it has an axis. Hence

$$\alpha^{-1}\eta\,\alpha\eta^{-1} = \iota \quad \text{or} \quad \eta\alpha = \alpha\eta.$$

EXERCISES

1 Prove the assertions (2)–(6) on bijections.

2 Show that isomorphic affine planes have isomorphic collineation groups.

3 Show that any two affine planes of order three are isomorphic.

4 Using Chapter 1, Exercise 8, prove that the group of the collineations of the plane of order two is isomorphic to the group of the permutations of a set of four elements. Each of the three groups $T(\Pi)$ has the order two and T is the non-cyclic group of order four.

5 Let l be a trace of the collineation α. If l is either parallel to a fixed line or incident with a fixed point of α, l is fixed.

6 Let τ be a translation in a finite affine plane. Then the order of τ divides the order of the plane.

7 Suppose τ and τ' are two translations $\neq \iota$ and in distinct directions. Show by a geometric argument that $\tau\tau' = \tau'\tau$; cf. Theorem 19 and the second proof of Theorem 16.

8 Let δ be a dilatation with the centre O and let τ be a translation; $P\delta = Q \neq P$, $O\tau = R \neq O$. (i) Construct the centre of the dilatation $\tau\delta$. (ii)* With δ, $\tau\delta$ will have the order n.

9 Suppose the fixed set $\mathbf{F}(\alpha)$ of the collineation α is a parallel pencil. Show that α is a translation in the direction of that pencil.

10 Every collineation with a centre is a dilatation.

11 Show that an involutory collineation with not more than one fixed point is a homothety. Thus the reflections in points are the involutory collineations with exactly one fixed point.

EXERCISES

12 Assume there are reflections in points and translations different from the identity. Show that the translations together with the reflections in points form a non-abelian normal subgroup N of the group of all the collineations. The translation group is a normal subgroup of index two of N. Its coset $N \backslash T$ consists of the reflections in points; cf. Exercise 13.

13 (i) Prove $\sigma\tau\sigma = \tau^{-1}$ for every translation τ and every reflection σ in a point.
(ii) Deduce: if an affine plane has a reflection in a point, its translation group is abelian.

14* Consider an affine plane with both a dilatation and a translation of the same order p; cf. Theorem 24(v). (i) Show that $p = 3$ is impossible. (ii) Let $p > 3$. Show that the group H is infinite and that T is non-abelian; cf. Theorems 9 and 19.

15 Let $D(O)$ be linearly transitive; cf. p. 32. Suppose there is a translation $\tau \in T(\Pi)$, $\tau \neq \iota$. (i) Show that
$$T(\Pi) = \{\delta^{-1}\tau\delta \mid \delta \in D(O)\}$$
and that $T(\Pi)$ is linearly transitive. (ii) If ord $\tau = n$, every translation $\neq \iota$ has the order n. (iii) If ord $\tau > 2$, there exists one and only one reflection in O [Hint: by (i) there is a $\delta \in D(O)$ such that $\tau^{-1} = \delta^{-1}\tau\delta$. Thus $\delta^2 = (\tau\delta)^2$]; cf. Exercise 13.

16 Give two distinct proofs of Theorem 28.

17 Suppose the fixed set of the collineation α contains a parallel pencil and a point Then α is an axial affinity.

18 Let α and α' be affinities with the same axis but different traces [thus $\alpha \neq \iota \neq \alpha'$]. Show that $\alpha\alpha' \neq \alpha'\alpha$.

19 Let α and α' be affinities with the distinct axes a and a', respectively. Then $\alpha\alpha' = \alpha'\alpha$ if and only if a and a' are traces of α' and α, respectively.

20* Every affine plane has a finite or countable subplane [a set is *countable* if it can be mapped bijectively onto the set of the positive integers].

3
Translation planes

3.1
LINEAR TRANSITIVITY

In the next three chapters we shall study families of affine planes which will be more and more restricted. The conditions which we shall impose can be expressed either as postulates for the existence of certain collineations or as configuration axioms.

Let G denote one of the groups T, $T(\Pi)$, $D(O)$, $H(a)$, or $A(a)$. If the points P and Q are distinct, then by Chapter 2 there is not more than one collineation in G which maps P onto Q. We call G *linearly transitive* if such a collineation actually exists for every pair P, Q for which this is not trivially impossible. Thus:

T is linearly transitive if it is *transitive*, i.e. if there is a translation mapping P onto Q for every pair P, Q.

$T(\Pi)$ is linearly transitive if such a translation exists for any pair P, Q such that there is a line in Π incident with P and Q.

$D(O)$ is linearly transitive if there is a dilatation with the centre O mapping P onto Q for any pair $P \neq O$, $Q \neq O$ such that O, P, Q are collinear.

$H(a)$ is linearly transitive if there is a homothety η with the trace a mapping P onto Q for any pair P, Q such that either P, $Q \mathrel{\text{I}} a$ or P, $Q \mathrel{\not{\text{I}}} a$, [$\eta$ was unique if P, $Q \mathrel{\not{\text{I}}} a$].

Finally, $A(a)$ is linearly transitive if given any pair P, Q such that P, $Q \mathrel{\not{\text{I}}} a$, there exists an affinity with the axis a which maps P onto Q.

3.2
THE CONFIGURATIONS OF TRANSLATION PLANES

In this chapter we study *translation planes*, i.e. planes $\mathfrak{A} = (\mathbf{P}, \mathbf{L}, \mathbf{I})$ whose translation groups T are linearly transitive. We formulate at once our first geometric statement.

STATEMENT (d) ['Little Affine Theorem of Desargues'].* *Suppose the three lines l_0, l_1, l_2 are parallel and mutually distinct. Let*

---

* It is named after Desargues (1593–1662), a civil engineer and architect who fathered a distinguished school of projective geometry.

3.2 THE CONFIGURATIONS OF TRANSLATION PLANES

$P_k, Q_k \, I \, l_k, \quad k = 0, 1, 2.$

Assume

$[P_0, P_1] \parallel [Q_0, Q_1], [P_0, P_2] \parallel [Q_0, Q_2].$

Then (1)

$[P_1, P_2] \parallel [Q_1, Q_2];$ (2)

cf. Figure 3.1.

REMARK 1.1 *Statement* (d) *holds in any affine plane if either* P_0, P_1, P_2 *or* Q_0, Q_1, Q_2 *are collinear or if* $P_k = Q_k$ *for some* k.

PROOF Our assertion is symmetric both in the indices 1 and 2 and in the P's and Q's. Thus it is sufficient to consider only the following three cases.
(i) Assume P_0, P_1, P_2 are collinear. Then

$[Q_0, Q_2] \parallel [P_0, P_2] = [P_0, P_1] \parallel [Q_0, Q_1].$

Hence

$[Q_0, Q_2] = [Q_0, Q_1] = [Q_1, Q_2],$

and therefore

$[Q_1, Q_2] = [Q_0, Q_2] \parallel [P_0, P_2] = [P_1, P_2].$

(ii) Assume $P_0 = Q_0$. Then

$[P_0, P_k] \parallel [Q_0, Q_k] = [P_0, P_k].$

Thus $Q_k = [l_k, [P_0, P_k]] = P_k$ for $k = 1, 2$; cf. Chapter 1, Theorem 5. Hence $[P_1, P_2] = [Q_1, Q_2]$.

(iii) The case $P_1 = Q_1$ is similar. □

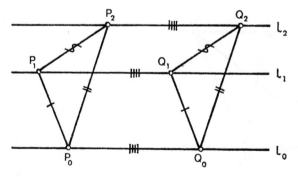

Figure 3.1

THEOREM 1 *The little affine theorem* (d) *of Desargues holds true in a translation plane.*

PROOF There is a translation τ such that

$$P_0\tau = Q_0. \tag{3}$$

In the second proof of Chapter 2, Theorem 16, we obtained the point $P_k\tau$ as the intersection of the trace l_k of P_k with the line $[P_0, P_k]\tau = [Q_0, P_k\tau]$ through Q_0 parallel to $[P_0, P_k]$. Thus $P_k\tau = Q_k$; $k = 1, 2$. This yields

$$[P_1, P_2] \parallel [P_1, P_2]\tau = [P_1\tau, P_2\tau] = [Q_1, Q_2].\square$$

We obtain Statement (d)$_\Pi$ from (d) by making the additional assumption that the lines l_k belong to the parallel pencil Π. Since we needed for our proof merely the existence of a translation τ which satisfied (3), we have

COROLLARY 1.2 *If $T(\Pi)$ is linearly transitive, then* (d)$_\Pi$ *holds true.*

We wish to motivate a second statement. Let $\mathfrak{A}$ be a translation plane. By Chapter 2, Theorem 19, the groups T and $T(\Pi)$ are abelian. Let τ and τ' be two translations $\neq \iota$ in $T(\Pi)$ and let l_1, l_2 denote two distinct lines of Π;

$$A_1 \operatorname{I} l_1, \qquad C_2 \operatorname{I} l_2.$$

Put

$$B_1 = A_1\tau, \qquad C_1 = B_1\tau', \qquad B_2 = C_2\tau', \qquad A_2 = B_2\tau. \tag{4}$$

The first and last equations imply

$$[A_1, B_2] \parallel [B_1, A_2] \tag{5}_1$$

The two other equations yield

$$[B_1, C_2] \parallel [C_1, B_2]. \tag{5}_2$$

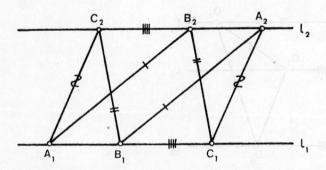

Figure 3.2

3.2 THE CONFIGURATIONS OF TRANSLATION PLANES

From (4) we obtain using the commutativity of $T(\Pi)$

$$C_1 = B_1\tau' = A_1\tau\tau' \quad \text{and} \quad A_2 = B_2\tau = C_2\tau'\tau = C_2\tau\tau'. \tag{6}$$

Hence

$$[A_1, C_2] \parallel [C_1, A_2]; \tag{7}$$

cf. Figure 3.2.

Conversely, assume $(5)_1$ and $(5)_2$. Then there are translations τ and τ' in $T(\Pi)$ which satisfy (4). This implies (6) and therefore (7).

We have shown that the following assertion is valid in a translation plane:

STATEMENT (p) ['Little Affine Theorem of Pappus'].* *Suppose the lines l_1 and l_2 are parallel and distinct. Let A_k, B_k, C_k be incident with l_k; $k = 1, 2$. Assume*

$$[A_1, B_2] \parallel [B_1, A_2], \quad [B_1, C_2] \parallel [C_1, B_2]. \tag{5}$$

Then

$$[A_1, C_2] \parallel [C_1, A_2]. \tag{7}$$

We repeat:

THEOREM 2 *Statement* (p) *holds in translation planes.*

Our argument has also yielded

COROLLARY 2.1 *If $T(\Pi)$ is linearly transitive and abelian in an affine plane $\mathfrak{A}$ then $(p)_\Pi$ holds true in $\mathfrak{A}$.* [We obtain $(p)_\Pi$ from (p) by adding the assumption that l_1 and l_2 belong to Π.]

A special case of Statement (p) applies to any affine plane.

REMARK 2.2 *Statement* (p) *always holds if A_1, B_1, C_1 or A_2, B_2, C_2 are not mutually distinct.*

The proof is analogous to that of Remark 1.1.

THEOREM 3 (d) *implies* (p).

PROOF (cf. Figure 3.3) On account of Remark 2.2 we may assume that the three points A_k, B_k, C_k are mutually distinct; $k = 1, 2$. Thus $A_1 \neq C_1$ and the lines $[A_1, B_2]$ and $[C_1, B_2]$ intersect. Hence the lines through C_1 parallel to $[A_1, B_2]$ and through A_1 parallel to $[C_1, B_2]$ also intersect, say at D.

Since $[D, A_1] \parallel [C_1, B_2] \parallel [B_1, C_2]$, we may apply Statement (d) to the triplets A_1, B_2, C_2 and D, C_1, B_1. Thus

$$[A_1, B_2] \parallel [D, C_1] \quad \text{and} \quad [B_2, C_2] \parallel [C_1, B_1]$$

implies $[A_1, C_2] \parallel [D, B_1]$.

---

* Pappus (approximately 320 AD) lectured at the Museum in Alexandria on mathematics, astronomy, geography, and the interpretation of dreams. His main work, the Synagogue [=collection], reports on and complements classical mathematics. He seems to have anticipated Desargues' Theorem.

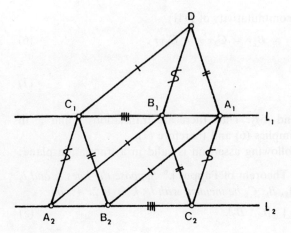

Figure 3.3

Symmetrically, we obtain

$[C_1, A_2] \parallel [D, B_1]$.

Hence $[A_1, C_2] \parallel [D, B_1] \parallel [C_1, A_2]$. □

Combining Theorems 1 and 3, we once more obtain Theorem 2.

We require a third statement. To motivate it we first assume that $\mathfrak{A}$ is a translation plane. Let τ be a translation $\neq \iota$ and let P_1 be any point. Using $Q_1 = P_1\tau$ we construct the point $Q_1' = P_1'\tau$ for a given point P_1' on $l_1 = [P_1, Q_1]$ in two different ways: choose $l_2 \parallel l_1$, $l_2 \neq l_1$. Let P_2 be any point on l_2. Construct the intersection Q_2 of l_2 with the line through Q_1 parallel to $[P_1, P_2]$; thus $Q_2 = P_2\tau$ and

$[P_1', P_2] \parallel [P_1'\tau, P_2\tau] = [Q_1', Q_2]$.

Replacing P_2 by any point P_2' on l_2 and repeating our construction, we arrive at a point $Q_2' = P_2'\tau$ and

$[P_1', P_2'] \parallel [Q_1', Q_2']$;

cf. Figure 3.4.

It is now easy to verify that the following theorem holds in a translation plane.

STATEMENT (s) ['Little Shear Theorem']: *Let $l_1 \parallel l_2$; $l_1 \neq l_2$. Let P_k, P_k', Q_k, Q_k' be incident with l_k ($k = 1, 2$) and let*

$$[P_1, P_2] \parallel [Q_1, Q_2], \quad [P_1, P_2'] \parallel [Q_1, Q_2'], \quad [P_1', P_2] \parallel [Q_1', Q_2]. \tag{8}$$

Then

$$[P_1', P_2'] \parallel [Q_1', Q_2']. \tag{9}$$

3.2 THE CONFIGURATIONS OF TRANSLATION PLANES

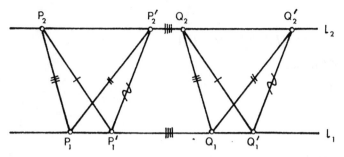

Figure 3.4

The preceding statement is symmetric both in the P's and Q's and in the indices 1 and 2. If, for example, $P_1 = Q_1$ (8) implies $P_2 = Q_2$, $P_2' = Q_2'$, $P_1' = Q_1'$ and hence (9). Similarly, (9) follows from (8) if $P_1' = Q_1'$ or if $P_1 = P_1'$. This yields

REMARK 4.1 *Statement* (s) *is valid in any affine plane unless*

$$P_k \neq Q_k, \quad P_k' \neq Q_k', \quad P_k \neq P_k', \quad Q_k \neq Q_k', \quad k = 1, 2. \tag{10}$$

We next prove

THEOREM 4 (p) *implies* (s).

PROOF (cf. Figure 3.5) Construct $P_2'' \, I \, l_2$ and $Q_1'' \, I \, l_1$ such that

$$[P_1, P_2''] \parallel [Q_1'', Q_2] \parallel [P_1', P_2']. \tag{11}$$

Since

$$[P_1, P_2] \parallel [Q_1, Q_2] \quad \text{and} \quad [P_1, P_2''] \parallel [Q_1'', Q_2],$$

Statement (p) yields

$$[Q_1, P_2''] \parallel [Q_1'', P_2].$$

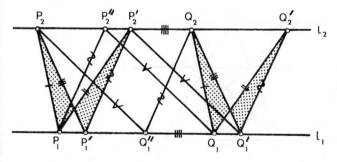

Figure 3.5

Furthermore

$[P_1', P_2'] \parallel [Q_1'', Q_2]$ and $[P_1', P_2] \parallel [Q_1', Q_2]$

imply on account of (p) that

$[Q_1'', P_2] \parallel [Q_1', P_2']$.

Hence

$[Q_1, P_2''] \parallel [Q_1', P_2']$.

Also $[P_1, P_2'] \parallel [Q_1, Q_2']$. Therefore by (p),

$[P_1, P_2''] \parallel [Q_1', Q_2']$.

Thus (11) implies (9). □

We obtain from our proof

COROLLARY 4.2 $(p)_\Pi$ *implies* $(s)_\Pi$.

[The statement $(s)_\Pi$ is obtained from (s) by adding the assumption that $\{l_1, l_2\} \subset \Pi$.]

By Theorems 3 and 4, (s) follows from (d). Since we could not deduce $(p)_\Pi$ from $(d)_\Pi$, the preceding results are not sufficient to prove

REMARK 4.3 $(d)_\Pi$ *implies* $(s)_\Pi$.

We give a direct proof assuming (10). If

$[P_1, P_2] \parallel [P_1', P_2']$ and $[Q_1, Q_2] \parallel [Q_1', Q_2']$,

$(s)_\Pi$ follows from $[P_1, P_2] \parallel [Q_1, Q_2]$. Thus we may assume that, e.g.,

$[P_1, P_2] \not\parallel [P_1', P_2']$;

cf. Figure 3.6.

Let P denote the intersections of these lines. Thus P is incident neither with l_1 nor with l_2. The line $l \in \Pi$ through P intersects $[Q_1, Q_2]$ at a point Q. Since

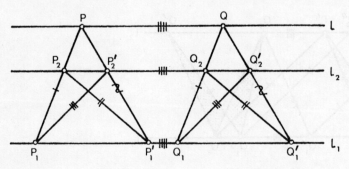

Figure 3.6

3.2 THE CONFIGURATIONS OF TRANSLATION PLANES

$[P_1, P] \parallel [Q_1, Q]$ and $[P_1, P_2'] \parallel [Q_1, Q_2']$,

$(d)_\Pi$ implies

$[P_2', P] \parallel [Q_2', Q]$.

Since

$[P_2, P] \parallel [Q_2, Q]$ and $[P_1', P_2] \parallel [Q_1', Q_2]$,

$(d)_\Pi$ yields

$[P_1', P] \parallel [Q_1', Q]$.   (12)

Thus

$[Q_1', Q] \parallel [P_1', P] = [P_2', P] \parallel [Q_2', Q]$

and hence

$[Q_1', Q] = [Q_2', Q]$.

The points Q_1', Q_2', Q are therefore collinear and (12) implies

$[Q_1', Q_2'] = [Q_1', Q] \parallel [P_1', P] = [P_1', P_2']$. □

The main result of this section is the following

THEOREM 5 *An affine plane is a translation plane if and only if the little affine theorem* (d) *of Desargues holds true.*

Because of Theorem 1 we may assume that the plane $\mathfrak{A} = (\mathbf{P}, \mathbf{L}, \mathbf{I})$ satisfies (d). Let P_0, Q_0 be two distinct points. We have to construct a translation τ such that

$P_0 \tau = Q_0$.   (13)

(i) We first construct a bijection τ of **P** onto itself which satisfies (13).

Put $l_0 = [P_0, Q_0]$. If $P \not{I} l_0$, let $Q = P\tau$ be the intersection of the lines parallel to l_0 through P and parallel to $[P_0, P]$ through Q_0; cf. Figure 3.7.

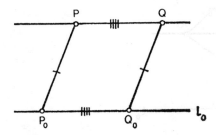

Figure 3.7

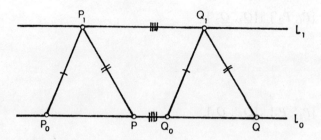

Figure 3.8

Before defining $Q = P\tau$ for points P on l_0, we choose any point P_1 outside l_0 and put $Q_1 = P_1\tau$. Then we repeat our construction for a point $P \, \mathrm{I} \, l_0$ replacing P_0 and Q_0 by P_1 and Q_1, respectively. This defines a map τ of $\mathbf{P}$ into itself, which, obviously, satisfies (13); cf. Figure 3.8.

In order to show that τ is a bijection, construct a second map τ' repeating our construction but interchanging the P's and Q's. Thus

$$P\tau = Q \Leftrightarrow Q\tau' = P. \tag{14}$$

Hence τ is a bijection with the inverse $\tau^{-1} = \tau'$.

(ii) We next prove

$$[A, B] \parallel [A\tau, B\tau] \tag{15}$$

for all the pairs of distinct points A, B. As this is trivial if $[A, B] \parallel l_0$, we may assume

$$[A, B] \not\parallel l_0.$$

First case (cf. Figure 3.9): $A, B \not\mathrel{\mathrm{I}} l_0$. By our construction

$[A, A\tau] \parallel [B, B\tau] \parallel l_0$,

$[P_0, A] \parallel [Q_0, A\tau]$, and $[P_0, B] \parallel [Q_0, B\tau]$.

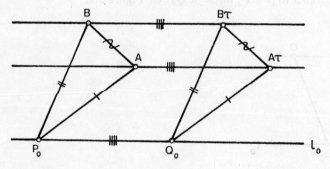

Figure 3.9

3.2 THE CONFIGURATIONS OF TRANSLATION PLANES

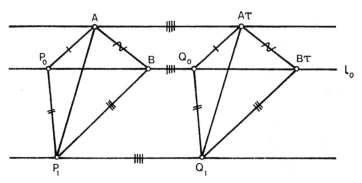

Figure 3.10

Thus (d) implies (15).

Second case (cf. Figure 3.10): $A \, \cancel{I} \, l_0$, $A \, \cancel{I} \, [P_1, Q_1]$, $B \, I \, l_0$. Then

$B\tau \, I \, l_0$, $[A, A\tau] \parallel [P_1, Q_1] \parallel l_0$,

and

$[P_0, A] \parallel [Q_0, A\tau]$, $[P_0, P_1] \parallel [Q_0, Q_1]$, $[P_1, B] \parallel [Q_1, B\tau]$.

Hence (d) applied to the triangles AP_0P_1 and $(A\tau)Q_0Q_1$ implies

$[A, P_1] \parallel [A\tau, Q_1]$.

Applying (d) once more to the triangles P_1AB and $Q_1(A\tau)(B\tau)$, we now obtain (15).

Third case: $A \, I \, [P_1, Q_1]$ and $B \, I \, l_0$. Then

$A\tau \, I \, [P_1, Q_1]$ and $B\tau \, I \, l_0$, $[P_1, Q_1] \parallel l_0$,

and

$[P_0, P_1] \parallel [Q_0, Q_1]$, $[P_0, A] \parallel [Q_0, A\tau]$, $[B, P_1] \parallel [B\tau, Q_1]$.

Thus (15) follows from (s), while (s) is a consequence of (d).
(iii) By Chapter 2, Theorem 8, (15) implies that the bijection τ of **P** onto itself can be completed to a homothety τ of $\mathfrak{A}$ in one and only one way. By our construction, the point mapping had no fixed points. Thus τ is a translation. It satisfied (13). □

By Remark 4.3, $(s)_\Pi$ follows from $(d)_\Pi$. Thus we obtain

COROLLARY 5.1 *Statement* $(d)_\Pi$ *holds true if and only if* $T(\Pi)$ *is linearly transitive.*

We collect the preceding results in two diagrams.
It is an unsolved problem whether (d) and (p) or (p) and (s) are equivalent. It can be shown that d_Π does not imply $(p)_\Pi$.

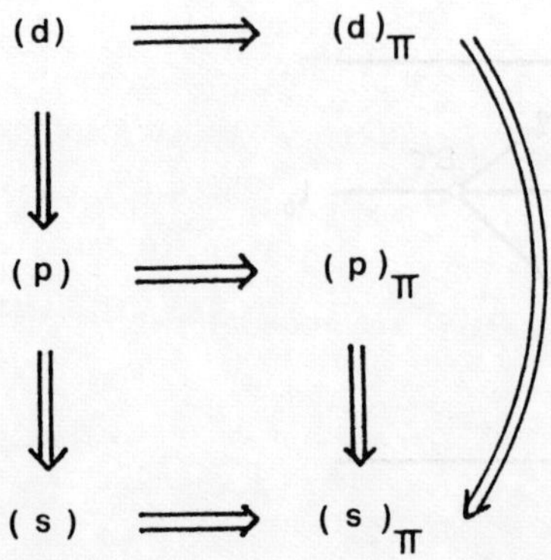

```
(d)  ⟹  (d)_Π
 ⇓        ⇓
(p)  ⟹  (p)_Π
 ⇓        ⇓
(s)  ⟹  (s)_Π
```

Diagram I

$\mathfrak{A}$ translation plane $\Leftrightarrow$ (d)
$T(\Pi)$ linearly transitive $\Leftrightarrow$ (d)$_\Pi$
$T(\Pi)$ lin. trans. and abelian $\Leftrightarrow$ (d)$_\Pi$ and (p)$_\Pi$.

Diagram II

In order to show that Statement (d) is an independent axiom which is not implied by Axioms A1–A3, we give an example of an affine plane which is not a translation plane. If a, b are real numbers, define

$$a \circ b = \begin{cases} 2ab & \text{if } b > 0, \\ ab & \text{otherwise.} \end{cases}$$

The *points* of our plane are the pairs of real numbers (a, b). The *lines* are the point sets

$$l(c) = \{(x, y) \mid x = c\} \quad \text{and} \quad l(m, d) = \{(x, y) \mid y = m \circ x + d\}. \tag{16}$$

A point is *incident* with a line l if it belongs to the point set l. Thus this plane is obtained from the real euclidean plane by replacing the lines of positive slope by lines broken along the y-axis; cf. Figure 3.11. The reader will verify that the axioms A1–A3 are satisfied and that

$$(x, y) \mapsto (x, y+a)$$

defines a translation parallel to the y-axis. However if Π denotes the pencil of the lines parallel to the x-axis, there are no translations $\neq \iota$ parallel to Π. [Also (s)$_\Pi$ does not hold; cf. the diagrams on this page and Figure 3.12.]

3.2 THE CONFIGURATIONS OF TRANSLATION PLANES

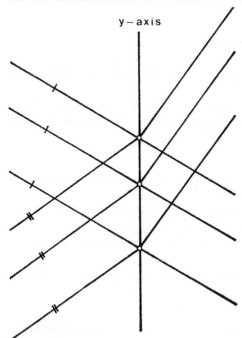

Figure 3.11

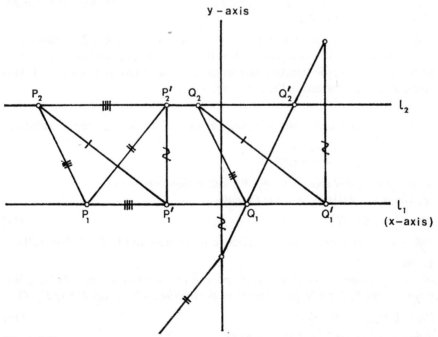

Figure 3.12

3.3*
THE PRIME KERNEL OF A TRANSLATION PLANE

In the second half of this chapter we study the group of the homotheties of a translation plane $\mathfrak{A}$.

By Chapter 2, Theorem 20, either every translation $\neq \iota$ has infinite order or there is a prime number p such that every translation $\neq \iota$ has the order p. In the first case, put $p = 0$. In either case, we call

$$p = \operatorname{char} \mathfrak{A}$$

the *characteristic* of $\mathfrak{A}$. Thus a translation plane of characteristic zero has no translations of finite order other than ι.

We remember that the *prime field* F_p *of characteristic* p is the rational field if $p = 0$ and the field with p elements if $p > 0$. *We now associate with each translation plane* $\mathfrak{A}$ *of characteristic* p *the prime field* F_p *of the same characteristic*. F_p *is called the prime kernel of* $\mathfrak{A}$.

Our results will be collected in the next three theorems. $\mathfrak{A}$ will always denote a translation plane; F_p will be its prime kernel.

THEOREM 6
(i) *Given any parallel pencil* Π *of* $\mathfrak{A}$, $T(\Pi)$ *contains a subgroup isomorphic to the additive group* F_{p+} *of* F_p.
(ii) *For every point* O *of* $\mathfrak{A}$, $D(O)$ *contains a subgroup isomorphic to the multiplicative group* $F_p{}^*$ *of* F_p.

By Theorem 6(ii), every translation plane of characteristic $\neq 2$ contains nontrivial dilatations. The two parts of this theorem illustrate the close connection between the geometric structure of a translation plane and its prime kernel. This connection is elaborated in the following theorems.

THEOREM 7
(i) *There exists exactly one mapping* $(\tau, r) \mapsto \tau^r$ *of* $T \times F_p$ *onto* T *satisfying* $\tau^1 = \tau$ *and*

$$\tau^{r+s} = \tau^r \tau^s \quad \text{and} \quad \tau^{rs} = (\tau^r)^s \tag{17}$$

for every $\tau \in T$ *and every* $r, s \in F_p$. *We call* τ^r *the rth power of* τ.
(ii) *Given* $\tau_0 \in T(\Pi)$, $\tau_0 \neq \iota$, *the formula*

$$r \mapsto \tau_0{}^r \quad \text{for all } r \in F_p \tag{18}$$

defines a monomorphism [i.e. an injective homomorphism] *of* $F_p{}^+$ *into* $T(\Pi)$.

THEOREM 8
(i) *Choose a point* O *of* $\mathfrak{A}$. *For every point* P *let* τ_P *denote the translation which maps* O *onto* P. *Let* r *be any element of the multiplicative group* $F_p{}^*$ *of* F_p. *Then*

$$P\delta_r = (O\tau_P{}^r) \quad \text{for all } P \tag{19}$$

defines a dilatation δ_r *with the centre* O.

3.3 THE PRIME KERNEL OF A TRANSLATION PLANE

(ii) *Through*

$$r \mapsto \delta_r \quad \text{for all } r \in F_p^*, \tag{20}$$

a monomorphism of F_p^* into $D(O)$ is defined. In particular

$$\delta_{rs} = \delta_r \delta_s = \delta_s \delta_r \quad \text{for all } r, s \in F_p^* \tag{21}$$

and

$$\delta_{r^{-1}} = \delta_r^{-1} \quad \text{for all } r \in F_p^*. \tag{22}$$

Theorem 6 being a corollary of Theorems 7(ii) and 8(ii), only Theorems 7 and 8 need to be proved. This proof will be prepared by two lemmas dealing with integral powers of translations.

Any translation τ generates the cyclic subgroup of T which consists of the integral powers of τ. Like every cyclic group it satisfies

$$\tau^{n+m} = \tau^n \tau^m \quad \text{and} \quad \tau^{nm} = (\tau^n)^m \tag{23}$$

for any integers n, m. Also, the group T being abelian, we have

$$\tau_1^n \tau_2^n = (\tau_1 \tau_2)^n \tag{24}$$

for every integer n and any translations τ_1 and τ_2.

From now on, let char $\mathfrak{A} = p$. Then

$$\tau^n = \iota \Leftrightarrow p \mid n, \tag{25}$$

if $\tau \neq \iota$ [$a \mid b$ means a divides b. Thus $0 \mid n \Leftrightarrow n = 0$].

LEMMA 7.1 *Let τ_1, τ_2 denote two translations $\neq \iota$ and in distinct directions. Then*

$$[P\tau_1^n, P\tau_2^n] \parallel [P\tau_1, P\tau_2] \tag{26}$$

for every point P and every integer n such that $p \nmid n$.

PROOF By our assumptions $\tau_1^{-1}\tau_2 \neq \iota$ and thus $(\tau_1^{-1}\tau_2)^n = \tau_1^{-n}\tau_2^n \neq \iota$; cf. (24) and (25). We have

$$P\tau_1(\tau_1^{-1}\tau_2) = P\tau_2.$$

Hence $[P\tau_1, P\tau_2]$ is a trace of $\tau_1^{-1}\tau_2$. Thus it also is a trace of $(\tau_1^{-1}\tau_2)^n = \tau_1^{-n}\tau_2^n$. Similarly

$$(P\tau_1^n)(\tau_1^{-n}\tau_2^n) = P\tau_2^n$$

implies that $[P\tau_1^n, P\tau_2^n]$ is a trace of $\tau_1^{-n}\tau_2^n$. Any two traces of this translation being parallel, we obtain (26). □

LEMMA 7.2 *Let $\tau_0 \in T$, $p \nmid n$. Then the equation*

$$\tau^n = \tau_0 \tag{27}$$

has one and only one solution $\tau \in T$.

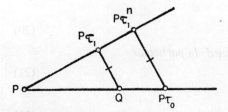

Figure 3.13

PROOF (cf. Figure 3.13) Choose any point P and any translation $\tau_1 \neq \iota$ which is not parallel to τ_0. The line through $P\tau_1$ parallel to $[P\tau_1{}^n, P\tau_0]$ intersects $[P, P\tau_0]$ at a point Q. Construct the translation τ which maps P onto Q. By Lemma 7.1,

$$[P\tau_1{}^n, P\tau^n] \parallel [P\tau_1, P\tau] = [P\tau_1, Q] \parallel [P\tau_1{}^n, P\tau_0].$$

Thus

$$[P\tau_1{}^n, P\tau^n] = [P\tau_1{}^n, P\tau_0].$$

Since $P\tau^n$ and $P\tau_0$ are both incident with $[P, Q] = [P, P\tau_0]$, this yields $P\tau^n = P\tau_0$. As there is only one translation which maps P onto $P\tau_0$, we obtain $\tau^n = \tau_0$.

If $\tau^n = \tau'^n$, (24) implies

$$(\tau^{-1}\tau')^n = \tau^{-n}\tau'^n = \iota.$$

Since $p \nmid n$, this yields $\tau^{-1}\tau' = \iota$ or $\tau' = \tau$. Thus equation (27) has exactly one solution τ. □

PROOF OF THEOREM 7 To (i). We first define τ^r for every $r \in F_p$. If $p > 0$, the prime field F_p can be identified with the field of the residue classes of the integers modulo p. Put

$$\bar{n} = \{m \mid m \equiv n \pmod{p}\}.$$

By (25), $\tau^{\bar{n}}$ is uniquely defined for every integer n by

$$\tau^{\bar{n}} = \tau^n. \tag{28}$$

This defines τ^r for every $r \in F_p$ and every $\tau \in T$. In this case, (17) follows at once from (28) and (23).

Let $p = 0$. Thus F_p is the rational field. Let $r = m/n$ denote any rational number. By Lemma 7.2, there is to every τ one and only one τ_0 such that $\tau_0{}^n = \tau^m$. We therefore try to define

$$\tau^{m/n} = \tau_0 \Leftrightarrow \tau^m = \tau_0{}^n. \tag{29}$$

This definition is justified if it is independent of the representation of m/n as a quotient of two integers.

3.3 THE PRIME KERNEL OF A TRANSLATION PLANE 47

Let $m/n = m'/n'$, Then $mn' = m'n$ and

$$\tau^m = \tau_0{}^n \Leftrightarrow \tau^{mm'} = \tau_0{}^{nm'} = \tau_0{}^{mn'} \Leftrightarrow \tau^{m'} = \tau_0{}^{n'}.$$

This equivalence now permits us the definition (29). Thus τ^r is defined for every $r \in F_0$ and every $\tau \in T$. It remains to verify (17) for $p = 0$.

Let $r = m/n$ and $s = m'/n'$ denote rational numbers. Put

$$\tau_0 = \tau^{(m/n)+(m'/n')}, \quad \tau_1 = \tau^{m/n}, \quad \tau_2 = \tau^{m'/n'}.$$

Thus

$$\tau^{mn'+m'n} = \tau_0{}^{nn'}, \quad \tau^m = \tau_1{}^n, \quad \tau^{m'} = \tau_2{}^{n'}.$$

Hence

$$(\tau_1\tau_2)^{nn'} = (\tau_1{}^n)^{n'}(\tau_2{}^{n'})^n = \tau^{mn'}\tau^{m'n} = \tau^{mn'+m'n} = \tau_0{}^{nn'},$$

and therefore, by Lemma 7.2, $\tau_1\tau_2 = \tau_0$. Thus

$$\tau^{(m/n)+(m'/n')} = \tau_0 = \tau_1\tau_2 = \tau^{m/n} \cdot \tau^{m'n'} \qquad \text{for any } \tau \in T.$$

For the proof of the second half of (17), write

$$\tau_0 = \tau^{m/n \cdot m'/n'}, \quad \tau_1 = \tau^{m/n}, \quad \tau_2 = \tau_1{}^{m'/n'}.$$

Thus

$$\tau^{mm'} = \tau_0{}^{nn'}, \quad \tau^m = \tau_1{}^n, \quad \tau_1{}^{m'} = \tau_2{}^{n'}.$$

Hence

$$\tau_2{}^{n'n} = (\tau_2{}^{n'})^n = (\tau_1{}^{m'})^n = (\tau_1{}^n)^{m'} = (\tau^m)^{m'} = \tau_0{}^{nn'} = \tau_0{}^{n'n},$$

and so $\tau_2 = \tau_0$ and

$$\tau^{m/n \cdot m'/n'} = \tau_0 = \tau_2 = \tau_1{}^{m'/n'} = (\tau^{m/n})^{m'/n'}$$

for all $\tau \in T$. This proves the first part of Theorem 7.

To (ii). By (17), (18) defines a homomorphism of $F_p{}^+$ into the group $T(\Pi)$ of the translations parallel to τ_0. We have to verify that it is injective.

Suppose r lies in the kernel of this homomorphism. If $p \neq 0$, then $r = \bar{n}$ for some integer n, and

$$\iota = \tau_0{}^r = \tau_0{}^{\bar{n}} = \tau_0{}^n$$

implies $p \mid n$ or $r = 0$; cf. (27).

If $p = 0$, $r = m/n$, then $\tau^r = \tau^{m/n} = \iota$ implies by (29) that $\tau^m = \iota^n = \iota$ and hence $m = 0$ and again $r = 0$. Thus the kernel of the homomorphism (18) is always equal to $\{0\}$. □

Before proving Theorem 8, we extend Lemma 7.1.

LEMMA 8.1 *Suppose the translations τ_1, τ_2, ι are mutually distinct. Let $r \in F_p^*$. Then*

$$[P\tau_1^r, P\tau_2^r] \parallel [P\tau_1, P\tau_2] \quad \text{for all } P.$$

PROOF Our assertion being trivial if τ_1 and τ_2 have the same direction, we may assume that their directions are distinct.

(i) Let $p > 0$. Since every $r \subset F_p^*$ is equal to a residue class n with $p \nmid n$, our statement follows from (28) and Lemma 7.1.

(ii) Assume $p = 0$. Let $r = m/n$ denote any rational number. Then there are translations τ_1', τ_2' such that $\tau_1^m = \tau_1'^n$ and $\tau_2^m = \tau_2'^n$. Hence by Lemma 7.1,

$$[P\tau_1^r, P\tau_2^r] = [P\tau_1', P\tau_2'] \parallel [P\tau_1'^n, P\tau_2'^n] = [P\tau_1^m, P\tau_2^m] \parallel [P\tau_1, P\tau_2]$$

for all P. □

PROOF OF THEOREM 8

To (i). Let $r \in F_p^*$. Obviously, δ_r is a map of the set of the points of $\mathfrak{A}$ into itself. We prove that δ_r is bijective by verifying that it has the inverse $\delta_{r^{-1}}$, i.e. that (22) holds true: We have

$$Q = P\delta_r = O\tau_P^r \Leftrightarrow \tau_Q = \tau_P^r \Leftrightarrow \tau_P = \tau_Q^{r^{-1}}$$

$$\Leftrightarrow P = O\tau_P = O\tau_Q^{r^{-1}} = Q\delta_{r^{-1}}.$$

This yields

$$Q = P\delta_r \Leftrightarrow P = Q\delta_{r^{-1}}.$$

By Lemma 8.1,

$$[P\delta_r, Q\delta_r] = [O\tau_P^r, O\tau_Q^r] \parallel [O\tau_P, O\tau_Q] = [P, Q].$$

Hence by Chapter 2, Theorem 8, the bijection δ_r of **P** can be extended to a homothety δ_r of $\mathfrak{A}$. As

$$O\delta_r = O\tau_0^r = O\iota^r = O,$$

this homothety is a dilatation with the centre O. This proves (i).

To (ii). Obviously, (20) defines a map of F_p^* into the group $D(O)$. By (19),

$$O\tau_{P\delta_r} = P\delta_r = O\tau_P^r, \quad \text{thus } \tau_{P\delta_r} = \tau_P^r.$$

Hence by (17),

$$P\delta_{rs} = O\tau_P^{rs} = O(\tau_P^r)^s = O\tau_{P\delta_r}^s = (P\delta_r)\delta_s = P\delta_r\delta_s$$

for every P. This yields (21). In particular, our map (20) is a homomorphism of F_p^* into $D(O)$.

It remains to prove that (20) is injective. Let r lie in its kernel. Thus $\delta_r = \iota$ and

$$O\tau_P = P = P\delta_r = O\tau_P^r.$$

Hence $\tau_P = \tau_P{}^r$ and $\tau_P{}^{r-1} = \tau_P{}^r \tau_P{}^{-1} = \iota$ for all P. Thus $r - 1 = 0$ or $r = 1$, by Theorem 7(ii), and (20) is a monomorphism. □

COROLLARY 8.2 *If $\mathfrak{A}$ is a translation plane of characteristic $p \neq 2$, there exists to every point O a reflection in O.*

PROOF By Theorem 8 (22), we have $\delta_{-1}{}^{-1} = \delta_{-1}$ or $\delta_{-1}{}^2 = \iota$. If $p \neq 2$, then $-1 \neq 1$ and we obtain

$$P\delta_{-1} = O\tau_P{}^{-1} \neq O\tau_P = P \quad \text{if } P \neq O.$$

Hence $\delta_{-1} \neq \iota$. Thus δ_{-1} is an involution. This proves our corollary. □

EXERCISES

1 Prove that Statement (p) is equivalent to the following Statement (p'): Let l_1 and l_2 be parallel and distinct; P_k, P_k', Q_k, $Q_k' I l_k$ for $k = 1, 2$;

$[P_1, P_2] \parallel [Q_1', Q_2']$, $[P_1, P_2'] \parallel [Q_1, Q_2']$, $[P_1', P_2] \parallel [Q_1', Q_2]$.

Then $[P_1', P_2'] \parallel [Q_1, Q_2]$; cf. Figure 3.14.

2 Let $\mathfrak{A}$ be any affine plane such that the diagonals of every parallelogram are parallel. Show that (p) holds true in $\mathfrak{A}$. If $\mathfrak{A}$ admits a translation $\tau \neq \iota$, what is the order of τ?

3 Show that (d) implies the following Statement (s'): Let l_0, l_1, l_2 be parallel and mutually distinct; $P_0, P_0', Q_0 Q_0' I l_0; P_1, Q_1 I l_1; P_2, Q_2 I l_2$. Assume $[P_0, P_1] \parallel [Q_0, Q_1]$, $[P_0', P_1] \parallel [Q_0', Q_1]$, $[P_0, P_2] \parallel [Q_0, Q_2]$. Then $[P_0', P_2] \parallel [Q_0', Q_2]$; cf. Figure 3.15.

4* Prove that (d) is equivalent to Statement (r): Let $l_1 = [P_1, Q_1] \parallel l_2 = [P_2, Q_2]$. Assume $[P_1, P_2] \parallel [Q_1, Q_2]$, $[P_1, P_3] \parallel [Q_2, Q_3]$, $[P_2, P_3] \parallel [Q_1, Q_3]$. Then $[Q_1, P_3] \parallel [P_2, Q_3]$; cf. Figure 3.16.

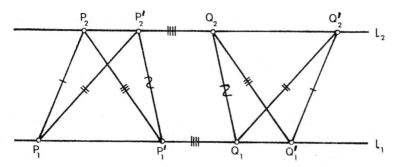

Figure 3.14

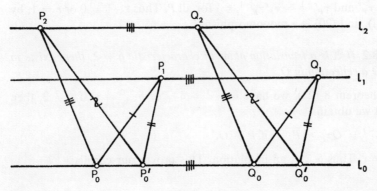

Figure 3.15

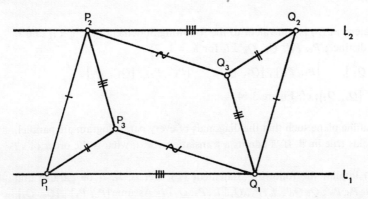

Figure 3.16

5 A set T of bijections of a set S onto itself is called [sharply simply] *transitive* if to every pair A, B in S there exists [exactly] one $\tau \in T$ such that $A\tau = B$. Prove: Suppose the set T is transitive and abelian. Then every bijection of S onto itself which commutes with every $\tau \in T$ belongs to T, and T is a sharply simply transitive group.

6 Let A be a translation plane. Prove that a collineation which commutes with every translation is itself a translation.

7 If $(p)_\Pi$ holds in an affine plane, $T(\Pi)$ is abelian.

8 Elaborate the example on p. 42.

9 Let F denote the field of order p^k; $p > 2$, $k > 1$. Let F^+ be the normal subgroup of F^* consisting of the squares of all the elements $\neq 0$. Define

$$x \circ y = \begin{cases} xy & \text{if } y \in F^+, \\ x^p y & \text{otherwise.} \end{cases}$$

EXERCISES

The points of $\mathfrak{A}$ are the ordered pairs (x, y) of elements of F; the lines are the point sets (16).
(i) Let $n \in F \setminus F^+$. Show that $x^p - nx = y^p - ny$ implies $x = y$ and deduce that the equation $x^p - nx = c$ has, for any $c \in F$, one and only one solution x.
(ii) Prove that $\mathfrak{A}$ satisfies the axioms A1–A3.
(iii) The group of the translations parallel to the y-axis is linearly transitive and abelian.
(iv) There are no translations $\neq \iota$ parallel to the x-axis.
(v) Deduce that there are at least two non-isomorphic planes of order p^k if $p > 2$, $k > 1$; cf. Chapter 1, Exercise 9 and Chapter 4, Exercise 7.

10 If an affine plane has reflections in each point it is a translation plane of characteristic $\neq 2$.

11* Prove Corollary 8.2 directly.

12 Let $r \in F_p{}^*$. (i) Verify $(\tau_1 \tau_2)^r = \tau_1{}^r \tau_2{}^r$ and show that the mapping $\tau \mapsto \tau^r$ defines an automorphism of T. (ii) Show that $\tau^r = \delta_r{}^{-1} \tau \delta_r$.

13 (i) Show that the group T of the translations of a translation plane $\mathfrak{A}$ is isomorphic to the additive group of a vector space over a prime field. (ii) Prove that this isomorphism induces a homomorphism of the group of the collineations of $\mathfrak{A}$ into the group of the linear transformations of the vector space; cf. Exercise 12(ii). Determine the kernel of this homomorphism. (iii) Let $\mathfrak{A}$ be finite. Show that the order of $\mathfrak{A}$ is a power of char $\mathfrak{A}$.

14 Let char $\mathfrak{A} \neq 2$. If $Q = P\tau$, the point $P\tau^{\frac{1}{2}}$ is called the *mid-point* between P and Q. (i) Show that this concept is symmetric in P and Q. (ii) Let $l \parallel h$. Prove that the locus of the mid-points of all the pairs P, Q with $P \operatorname{I} l$ and $Q \operatorname{I} h$ is a line parallel to both.

4
Desarguesian planes

4.1
THE DILATATION GROUPS $D(O)$

An affine plane $\mathfrak{A}$ is called *desarguesian* if the following statement holds true.

STATEMENT (D) ['Affine Theorem of Desargues']. *Let l_0, l_1, l_2 denote three mutually distinct lines with a common intersection O. The points P_k, Q_k are incident with l_k and distinct from O; $k = 0, 1, 2$. Assume*

$$[P_0, P_1] \parallel [Q_0, Q_1] \quad and \quad [P_0, P_2] \parallel [Q_0, Q_2]. \tag{1}$$

Then

$$[P_1, P_2] \parallel [Q_1, Q_2]; \tag{2}$$

cf. Figure 4.1.

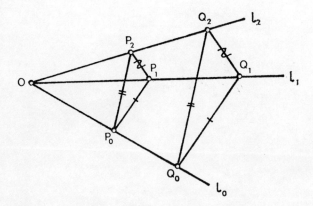

Figure 4.1

4.1 THE DILATATION GROUPS $D(O)$

REMARK 1.1 *Statement* (D) *holds true in any affine plane if either* P_0, P_1, P_2 *or* Q_0, Q_1, Q_2 *are collinear or if* $P_k = Q_k$ *for some k.*

PROOF See Chapter 3, Remark 1.1.

THEOREM 1 *If the dilatation group $D(O)$ is linearly transitive for every choice of O, the plane is desarguesian;* cf. Chapter 3, Theorem 1.

PROOF We use the above notation. By our assumption there exists a dilatation δ in $D(O)$ such that $P_0\delta = Q_0$. Since $h\delta \parallel h$ for every h and the traces of δ are the lines through O, $P_1\delta$ is the intersection of $[O, P_1]$ with the line through $P_0\delta = Q_0$ parallel to $[P_0, P_1]$. Thus $P_1\delta = Q_1$. Similarly $P_2\delta = Q_2$ and thus $[P_1, P_2] \parallel [P_1, P_2]\delta = [P_1\delta, P_2\delta] = [Q_1, Q_2]$. □

Let (D)$_O$ denote Statement (D) with the additional assumption that the lines l_0, l_1, l_2 are incident with the given point O. Our proof of Theorem 1 shows

COROLLARY 1.2 *If the group $D(O)$ is linearly transitive Statement* (D)$_O$ *will hold.*

Before studying the converse of Theorem 1, we prove

THEOREM 2 (D) *implies* (d).

COROLLARY 2.1 *Every desarguesian affine plane is a translation plane.*

PROOF OF THEOREM 2 Let l_0, l_1, l_2 denote three mutually distinct parallel lines. Suppose the points P_k and Q_k are incident with l_k ($k = 0, 1, 2$) and satisfy

$$[P_0, P_1] \parallel [Q_0, Q_1], \qquad [P_0, P_2] \parallel [Q_0, Q_2]. \tag{1}$$

We have to prove

$$[P_1, P_2] \parallel [Q_1, Q_2].$$

In our proof we may assume that $P_k \neq Q_k$ for $k = 0, 1, 2$ and that neither P_0, P_1, P_2 nor Q_0, Q_1, Q_2 are collinear.
Assume

$$[P_1, P_2] \not\parallel [Q_1, Q_2]. \tag{2'}$$

We first construct a configuration to which we shall later apply (D): Let h be the line through Q_1 parallel to $[P_1, P_2]$ and Q_2' the intersection of h and $[Q_0, Q_2]$. Put $l_2' = [P_2, Q_2']$, $O = [l_2', l_1]$, and $l_0' = [O, P_0]$. Finally, let Q_0' denote the intersection of l_0' and $[Q_0, Q_2]$; cf. Figure 4.2.

By our construction

$$[P_1, P_2] \parallel [Q_1, Q_2'] \quad \text{and} \quad [P_0, P_2] \parallel [Q_0', Q_2'].$$

Thus we may apply (D) to the triangles $P_0P_1P_2$ and $Q_0'Q_1Q_2'$ obtaining $[P_0, P_1] \parallel [Q_0', Q_1]$. Hence $[Q_0', Q_1] \parallel [Q_0, Q_1]$ and $Q_1 \text{ I } [Q_0', Q_1], [Q_0, Q_1]$.

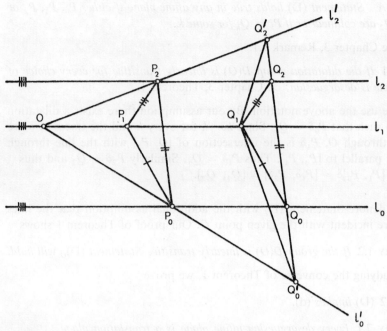

Figure 4.2

Thus $[Q_0', Q_1] = [Q_0, Q_1]$. This implies $Q_0' = Q_0$ and $l_0' = [P_0, Q_0] = l_0$. Since $O \text{ I } l_0, l_1$, and $l_0 \parallel l_1$, we obtain $l_0 = l_1$; contradiction. □

This proof is not complete: we have not discussed the existence and uniqueness of the elements of our construction and we have not made sure that the assumptions of (D) are satisfied. The details of such verifications are easy but cumbersome. It may be instructive to go through them at least once. In later proofs we shall tend to omit the corresponding discussions.

By our assumptions,

$$l_i \neq l_k \quad \text{if } i \neq k; i, k = 0, 1, 2; \tag{3}$$

$$P_k \neq Q_k, \quad k = 0, 1, 2; \tag{4}$$

$$l_i \neq [P_k, P_h], [Q_k, Q_h], \quad i, k, h = 0, 1, 2; k \neq h; \tag{5}$$

$$[P_i, P_k] \neq [P_i, P_h], \quad [Q_i, Q_k] \neq [Q_i, Q_h] \tag{6}$$

if $i, k, h = 0, 1, 2$; i, k, h mutually distinct.

I We first prove the existence of the elements of our construction:

(i) Q_2' exists: since $[Q_0, Q_2] \parallel [P_0, P_2] \not\parallel [P_1, P_2] \parallel h$, the lines $[Q_0, Q_2]$ and h must intersect.

4.1 THE DILATATION GROUPS $D(O)$

(ii) l_2' exists: suppose $P_2 = Q_2'$. Then $P_2, Q_2 \mathbin{I} l_2, [Q_0, Q_2]$, and $l_2 \neq [Q_0, Q_2]$ would imply $P_2 = Q_2$; cf. (5); contradiction.

(iii) $l_2' \nparallel l_1$ and thus O exists: suppose $l_2' \parallel l_1$. Then $l_2' \parallel l_2$. Since $P_2 \mathbin{I} l_2', l_2$, this would imply $l_2' = l_2$. As $Q_2, Q_2' \mathbin{I} l_2, [Q_0, Q_2]$, and $l_2 \neq [Q_0, Q_2]$, we would obtain $Q_2 = Q_2'$ and hence $[P_1, P_2] \parallel [Q_1, Q_2'] = [Q_1, Q_2]$, This contradicts (2').

(iv) l_0' exists: since l_0 and l_1 are distinct parallel lines, $P_0 \mathbin{I} l_0$ and $O \mathbin{I} l_1$ implies $P_0 \neq O$.

(v) Finally, Q_0' exists: suppose $l_0' \parallel [Q_0, Q_2]$. Then $l_0' \parallel [P_0, P_2]$. Since $P_0 \mathbin{I} l_0'$, $[P_0, P_2]$, this yields $l_0' = [P_0, P_2]$. Thus $O = [l_1, [P_0, P_2]]$. Since l_1 and l_2 are distinct parallel lines and $O \mathbin{I} l_1, P_2 \mathbin{I} l_2$, we have $O \neq P_2$. Hence $l_2' = [O, P_2] = [P_0, P_2]$. The lines $[P_0, P_2]$ and $[Q_0, Q_2]$ being parallel and having the point Q_2' in common, they would have to be identical. This is excluded by (4).

In order to verify that $P_0 P_1 P_2$ and $Q_0' Q_1 Q_2'$ are triangles which satisfy the assumptions of (D), we have to show:

II The lines l_0', l_1, l_2' are mutually distinct.

III The points $P_0, P_1, P_2, Q_0', Q_1, Q_2'$ are distinct from O.

To II

(vi) Since $P_0 \mathbin{I} l_0'$ and $P_0 \not\mathbin{I} l_1$, we have $l_0' \neq l_1$. Similarly $l_2' \neq l_1$.

(vii) $l_0' \neq l_2'$: suppose $l_0' = l_2'$. As P_0 and P_2 lie on the distinct parallel lines l_0 and l_2, we have $P_0 \neq P_2$. The lines $[P_0, P_2]$ and $l_0' = l_2'$ having the points P_0 and P_2 in common, we obtain $l_2' = [P_0, P_2]$. Repeating the argument of (v), we arrive at a contradiction.

To III

(viii) By (iv), we have $O \neq P_0$; (v) yielded $O \neq P_2$.

(ix) Suppose $O = P_1$. Then $O \neq P_2$ implies $l_2' = [O, P_2] = [P_1, P_2] \parallel h$. As $Q_2' \mathbin{I} l_2', h$, this would yield $l_2' = h$. Hence $P_1, Q_1 \mathbin{I} l_1, l_2'$. Since $l_1 \neq l_2'$, we would obtain $P_1 = Q_1$. This contradicts (4).

(x) Suppose $O = Q_1$. Then $l_2', h \mathbin{I} Q_1, Q_2'$. Thus either $Q_1 = Q_2'$ or $l_2' = h$. If $Q_1 = Q_2'$, this point would be incident with $[Q_0, Q_2]$. Thus $[Q_0, Q_1] = [Q_0, Q_2]$, contradicting (6). Suppose then $l_2' = h$. Then $[P_1, P_2] \parallel h = l_2'$ and $P_2 \mathbin{I} l_2', [P_1, P_2]$ would yield $[P_1, P_2] = l_2'$ and $O = [l_2', l_1] = P_1$. Hence $P_1 = Q_1$, contradicting (4).

(xi) $O = Q_2'$ would imply $O, Q_1 \mathbin{I} h, l_1$. Since $O \neq Q_1$ by (x), this would yield $l_1 = h \parallel [P_1, P_2]$. As $P_1 \mathbin{I} l_1, [P_1, P_2]$, we would obtain $[P_1, P_2] = l_1$, contradicting (v).

(xii) Suppose $O = Q_0'$. Since $O \neq Q_2'$, this would imply $l_2' = [O, Q_2'] = [Q_0', Q_2'] = [Q_0, Q_2] \parallel [P_0, P_2]$. The lines l_2' and $[P_0, P_2]$ having P_2 in common, we would obtain $[P_0, P_2] = l_2' = [Q_0, Q_2]$; in particular, $P_0 = [l_0, [P_0, P_2]] = [l_0, [Q_0, Q_2]] = Q_0$. This contradicts (4).

IV In conclusion, we elaborate the application of (D) in our proof: applying (D) to $P_0 P_1 P_2$ and $Q_0' Q_1 Q_2'$ we obtained $[Q_0', Q_1] = [Q_0, Q_1]$. Hence $Q_0, Q_0' \mathbin{I}$

$[Q_0, Q_1], [Q_0, Q_2]$. By (6), $[Q_0, Q_1] \neq [Q_0, Q_2]$; therefore $Q_0 = Q_0'$. Also $Q_0, P_0 \text{ I } l_0, l_0'$ and $Q_0 \neq P_0$, by (4). Hence $l_0 = l_0'$. Thus $O \text{ I } l_0, l_1$ while l_0 and l_1 are distinct parallel lines by (3); contradiction.

We can characterize the desarguesian planes:

THEOREM 3 *An affine plane is desarguesian if and only if the dilatation group $D(O)$ is linearly transitive for every choice of O.*

On account of Theorem 1, we have to prove: suppose the affine plane $\mathfrak{A} = (\mathbf{P}, \mathbf{L}, \text{I})$ is desarguesian. Then each group $D(O)$ is linearly transitive.

Suppose the points O, P_0, Q_0 are collinear and mutually distinct. We try to construct a dilatation δ with the centre O such that

$$P_0 \delta = Q_0; \tag{7}$$

cf. Figure 4.3.

Let $l_0 = [O, P_0] = [O, Q_0]$. Choose a point P_1 not incident with l_0 and put $l_1 = [O, P_1]$. We now define the mapping $\delta: \mathbf{P} \to \mathbf{P}$ as follows:
(i) If $P \not{\text{I}} \, l_0$, intersect $l = [O, P]$ with the line through Q_0 parallel to $[P_0, P]$. If Q is the intersection, put $P\delta = Q$. In particular, $Q_1 = P_1 \delta$ is the intersection of l_1 with the line through Q_0 parallel to $[P_0, P_1]$.
(ii) If $P_0' \text{ I } l_0$, let $Q_0' = P_0'\delta$ be the intersection of l_0 with the line through Q_1 parallel to $[P_1, P_0']$: in particular $O\delta = O$.

We have now defined a map δ of $\mathbf{P}$ into itself with the one fixed point O and

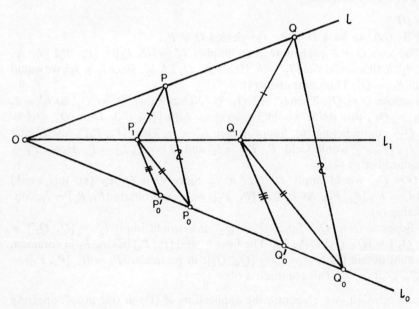

Figure 4.3

which satisfies (7). Interchanging the P's and Q's we obtain another map δ' of **P** into itself such that

$$P\delta = Q \Leftrightarrow Q\delta' = P.$$

Thus δ is a bijection with the inverse $\delta^{-1} = \delta'$.

Continuing our proof we next show

$$[P, P'] \parallel [P\delta, P'\delta] \qquad \text{for all } P \neq P'. \tag{8}$$

This is obvious if $O \mathrel{\text{I}} [P, P']$.

From now on we may assume that O, P, P' are not collinear. Put $P\delta = Q$, $P'\delta = Q'$.

Case (i) Neither P nor P' is incident with l_0. Then

$$[P_0, P] \parallel [Q_0, Q] \quad \text{and} \quad [P_0, P'] \parallel [Q_0, Q'].$$

Thus (8) follows from Statement (D).

Case (ii) From now on we may assume that, for example, $P' = P_0'$ is incident with l_0. Thus $P \mathrel{\bar{\text{I}}} l_0$. Put $Q_0' = P_0'\delta$.

(α) $P \mathrel{\bar{\text{I}}} l_1$. By Case (i), $[P_1, P] \parallel [Q_1, Q]$; and by our construction $[P_0', P_1] \parallel [Q_0', Q_1]$. Hence (D) yields

$$[P_0', P] \parallel [Q_0', Q].$$

(β) The point $P = P_1'$ is incident with l_1. By Chapter 1, Theorem 7, there is a line l_2 through O distinct from l_0 and l_1 and a point P_2 on l_2 distinct from O. Put $Q_1' = P_1'\delta$ and $Q_2 = P_2\delta$. By the preceding cases,

$$[P_1', P_2] \parallel [Q_1', Q_2] \quad \text{and} \quad [P_0', P_2] \parallel [Q_0', Q_2].$$

Hence by (D)

$$[P_0', P_1'] \parallel [Q_0', Q_1'].$$

This proves (8). We shall return to Case (ii) (β).

By Chapter 2, Theorem 8, (8) implies that our bijection δ can be completed to a homothety δ of $\mathfrak{A}$. As it has the fixed point O, this homothety belongs to $D(O)$. It satisfied (7). $\square$

Our proof and Corollary 1.2 imply

COROLLARY 3.1 *$D(O)$ is linearly transitive if and only if Statement $(D)_O$ holds true.*

4.2
THE SHEAR THEOREM

We formulate

STATEMENT (S) ['Shear Theorem']. *Let $l_1 \neq l_2$. Suppose P_1, P_1', Q_1, Q_1' are* incident with l_1 but not with l_2, and P_2, P_2', Q_2, Q_2' are incident with l_2 but not with l_1. Assume

$$[P_1, P_2] \parallel [Q_1, Q_2], \qquad [P_1, P_2'] \parallel [Q_1, Q_2'], \qquad [P_1', P_2] \parallel [Q_1', Q_2].$$

Then

$$[P_1', P_2'] \parallel [Q_1', Q_2'];$$

cf. Figure 4.4.

REMARK 4.1 *Statement* (S) *holds in any office plane unless*

$$P_k \neq P_k', \quad Q_k \neq Q_k', \quad P_k \neq Q_k, \quad P_k' \neq Q_k' \qquad \text{for } k = 1, 2; \tag{9}$$

cf. Chapter 3, Remark 4.1.

REMARK 4.2 *For $l_1 \parallel l_2$ we obtain the Little Shear Theorem* (s). *By Chapter 3, Theorems 3 and 4, it follows from* (d). *Hence* (D) *implies* (s) *by Theorem 2.*

If l_1 and l_2 are not parallel, we come back to case (ii)(β) of the proof of (8). This yields

THEOREM 4 (D) *implies* (S).

Let $(S)_O$ denote Statement (S) with the additional assumption that l_1 and l_2 are incident with the given point O. Then the proof of (8) yields

COROLLARY 4.3 $(D)_O$ *implies* $(S)_O$.

We next wish to prove the rather surprising result that the shear theorem (S) implies the affine theorem of Desargues (D).

Assume then that (S) holds true in our affine plane. The lines and points l_k, O, P_k, Q_k satisfy the assumptions of Statement (D). We wish to prove formula (2). We may assume that $P_k \neq Q_k$ and that neither P_0, P_1, P_2 nor Q_0, Q_1, Q_2 are collinear; $k = 0, 1, 2$.

(i) We first prove that, if (2) is false, the intersection R_2 of $[P_1, P_2]$ and $[Q_1, Q_2]$ is incident with l_0; cf. Figure 4.5.

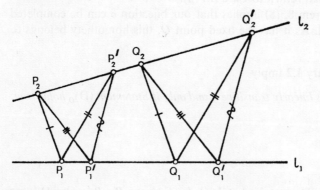

Figure 4.4

4.2 THE SHEAR THEOREM

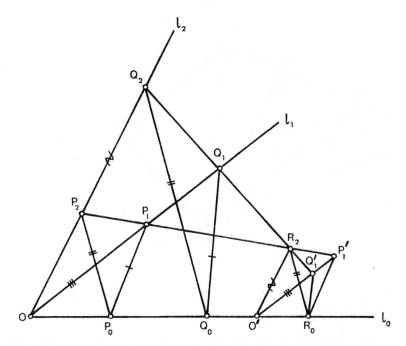

Figure 4.5

Assume $R_2 \not\mathrel{I} l_0$. By its definition, R_2 is not incident with l_1 or l_2. The line through R_2 parallel to l_2 intersects l_0 at some point O' and the line through R_2 parallel to $[P_0, P_2]$ and $[Q_0, Q_2]$ intersects l_0 at a point R_0. Finally, let P_1' and Q_1' denote the intersections of $[P_1, P_2]$ and $[Q_1, Q_2]$, respectively, with the line through O parallel to l_1.

Since

$$[O, P_1] \parallel [O', P_1'], \quad [O, P_2] \parallel [O', R_2], \quad [P_0, P_2] \parallel [R_0, R_2],$$

Statement (S) yields

$$[P_0, P_1] \parallel [R_0, P_1'].$$

Symmetrically, we obtain $[Q_0, Q_1] \parallel [R_0, Q_1']$. Hence

$$[R_0, P_1'] \parallel [P_0, P_1] \parallel [Q_0, Q_1] \parallel [R_0, Q_1']$$

and

$$[R_0, P_1'] = [R_0, Q_1'].$$

As the points P_1' and Q_1' are the intersections of this line with the line through O' parallel to l_1, this yields $P_1' = Q_1'$. This point is incident with both $[P_1, P_2]$ and $[Q_1, Q_2]$. Hence $P_1' = Q_1' = R_2$. Thus we would obtain

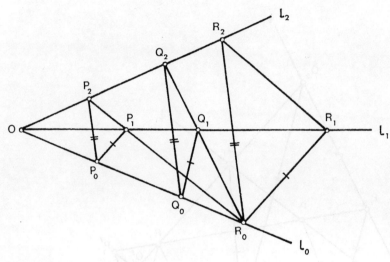

Figure 4.6

$[P_0, P_2] \parallel [R_0, R_2] = [R_0, P_1'] \parallel [P_0, P_1]$.

Hence $[P_0, P_2] = [P_0, P_1]$ and the points P_0, P_1, P_2 would be collinear; contradiction.

(ii) We can now prove that (S) implies (D); cf. Figure 4.6. Assume that (2) is false. Then by (i), the lines $[P_1, P_2]$ and $[Q_1, Q_2]$ intersect at a point R_0 incident with l_0. The line through R_0 parallel to $[P_0, P_k]$ intersects l_k at a point R_k. Thus

$[R_0, R_k] \parallel [P_0, P_k] \parallel [Q_0, Q_k], \quad k = 1, 2.$

Since P_0, P_1, P_2 are not collinear, R_0, R_1, R_2 cannot be collinear either. Thus $R_0 \not{I} [R_1, R_2]$.

If $[R_1, R_2]$ and $[P_1, P_2]$ were not parallel, their intersection would be incident with l_0 by (i). Thus it would be equal to R_0. As this has been excluded, we obtain

$[P_1, P_2] \parallel [R_1, R_2]$

and symmetrically

$[Q_1, Q_2] \parallel [R_1, R_2]$.

Thus we arrive at formula (2) after all; contradiction. □

Combining our result with Theorem 4, we obtain

THEOREM 5 *The affine theorem of Desargues* (D) *and the Shear Theorem* (S) *are equivalent.*

4.3
THE LINEAR TRANSITIVITY OF THE GROUPS $A(a)$

Our goal is the following result.

THEOREM 6 *The affine plane $\mathfrak{A}$ is desarguesian if and only if the group $A(a)$ is linearly transitive for every choice of the axis a.*

On account of Theorem 3, this assertion is a corollary of the following theorem.

THEOREM 7 *Let $\mathfrak{A} = (\mathbf{P}, \mathbf{L}, \mathbf{I})$ be any affine plane. Let a be any line in $\mathfrak{A}$. Then the following statements are equivalent:*
(i) *For every $O \mathbin{I} a$, the group $D(O)$ is linearly transitive.*
(ii) *The group $H(a)$ of the homotheties with the fixed line a is linearly transitive.*
(iii) *The group $A(a)$ of the affinities with the axis a is linearly transitive; cf. p. 32.*

PROOF OF THEOREM 7 We exclude the trivial case of the plane of order two.

(i) *implies* (ii): Let P, Q be two points not incident with a. Choose any point $O \mathbin{I} a$. Then there exists a point $O' \neq O$ on a such that $[O, P] \nparallel [O', Q]$. Let R be the intersection of $[O, P]$ and $[O', Q]$. Then $R \mathbin{\not{I}} a$ and there are dilatations $\delta \in D(O)$ and $\delta' \in D(O')$ such that $P\delta = R$ and $R\delta' = Q$. Hence $\eta = \delta\delta' \in H(a)$ and $P\eta = Q$.

If $P, Q \mathbin{I} a$, choose any point O on a distinct from P and Q. Then there is a $\delta \in D(O)$ such that $P\delta = Q$.

(ii) *implies* (iii): Given two points P_0 and Q_0 not incident with a, we have to construct an affinity α with the axis a such that

$$P_0 \alpha = Q_0; \tag{10}$$

cf. Figure 4.7.

By our assumption (ii) and by Chapter 2, Theorem 14, there is to any pair of points P, Q not incident with a, one and only one homothety in $H(a)$, say η_{PQ} which maps P onto Q. Thus

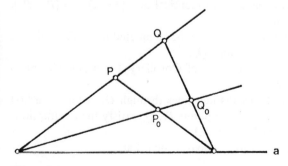

Figure 4.7

$P\eta_{PQ} = Q, \qquad \eta_{PQ} \in H(a)$.

Define the map α of **P** into itself by

$$P\alpha = Q_0 \eta_{P_0 P} \quad \text{if } P \not\mathrel{I} a, \tag{11}$$
$$P\alpha = P \quad \text{if } P \mathrel{I} a.$$

Interchanging the P's and Q's, we obtain a second map α' of **P** into itself. By our definition

$$P\alpha = Q \Leftrightarrow Q\alpha' = P \qquad \text{for all } P.$$

Thus α is a bijection of **P** onto itself with the inverse $\alpha^{-1} = \alpha'$.

Before proving that α preserves collinearity, we show

(*) *If* $P \neq Q$; $P, Q \not\mathrel{I} a$, *then the lines* $[P, Q], [P\alpha, Q\alpha], a$

belong to the same pencil.

As the homothety $\eta_{P_0 P}^{-1} \eta_{P_0 Q}$ lies in $H(a)$ and maps P onto Q, we have

$$\eta_{P_0 P}^{-1} \eta_{P_0 Q} = \eta_{PQ}.$$

Hence

$$(P\alpha)\eta_{PQ} = (Q_0 \eta_{P_0 P})(\eta_{P_0 P}^{-1} \eta_{P_0 Q}) = Q_0 \eta_{P_0 Q} = Q\alpha.$$

Thus the lines $[P, Q], [P\alpha, Q\alpha]$ and the fixed line a are traces of the same homothety η_{PQ}. In particular they belong to the same pencil. This proves (*).

Let P, Q, R be three mutually distinct collinear points. We wish to show that $P\alpha, Q\alpha, R\alpha$ are collinear. We may assume that not more than one of the points P, Q, R is incident with a.

If $R \mathrel{I} a$, then $R = [a, [P, Q]]$. Hence by (*), $R \mathrel{I} [P\alpha, Q\alpha]$. Thus $P\alpha, Q\alpha, R = R\alpha$ are collinear.

If none of the three points P, Q, R is incident with a, then (*) implies that $[P, Q], [P\alpha, Q\alpha], a$ and similarly $[P, R], [P\alpha, R\alpha], a$ belong to the same pencils. As $[P, Q] = [P, R] \neq a$, these two pencils are identical and $[P\alpha, Q\alpha] = [P\alpha, R\alpha]$. Thus $P\alpha, Q\alpha, R\alpha$ are again collinear.

By Chapter 2, Theorem 3, our bijection α can be completed to a collineation α. By (11) it has the axis a and maps P_0 onto Q_0.

(iii) *implies* (i): Assume (iii). On account of Corollary 3.1, it is sufficient to prove Statement (D)$_O$ for every $O \mathrel{I} a$.

Let l_0, l_1, l_2 denote three mutually distinct lines through O. We assume first that they are distinct from a. Suppose the points P_k, Q_k satisfy the assumptions of Statement (D)$_O$; $k = 0, 1, 2$; cf. Figure 4.1.

There are two affinities α and β with the axis a such that

$$P_0 \alpha = P_1, \qquad P_1 \beta = P_2.$$

4.3 THE LINEAR TRANSITIVITY OF THE GROUPS $A(a)$

If we can show that

$$Q_1\beta = Q_2, \qquad (12)$$

both $[P_1, P_2]$ and $[Q_1, Q_2]$ will be traces of β; thus they will be parallel.

We have

$$l_0\alpha = [O, P_0]\alpha = [O\alpha, P_0\alpha] = [O, P_1] = l_1.$$

Thus $Q_0 \, \mathrm{I} \, l_0$ implies $Q_0\alpha \, \mathrm{I} \, l_1$. The traces of α being parallel, $Q_0\alpha$ must be the intersection of l_1 with the line $[Q_0, Q_1]$ through Q_0 parallel to $[P_0, P_1]$, i.e.

$$Q_1 = Q_0\alpha.$$

Since $\alpha\beta$ also is an affinity with the axis a, we obtain in the same way $Q_2 = Q_0\alpha\beta$. This yields (12).

If $a = l_0$, apply only β. The points P_0 and Q_0 now being fixed, (1) implies

$$[Q_0, Q_2] \parallel [P_0, P_2] = [P_0, P_1]\beta \parallel [Q_0, Q_1]\beta = [Q_0, Q_1\beta].$$

Since $Q_1\beta \, \mathrm{I} \, l_1\beta = l_2$, we arrive again at (12).

If $a = l_2$, we use only α. Since P_2 and Q_2 are fixed, we have

$$[Q_1, Q_2] = [Q_0\alpha, Q_2\alpha] = [Q_0, Q_2]\alpha \parallel [P_0, P_2]\alpha = [P_1, P_2].$$

The case $a = l_1$ is symmetric to the case $a = l_2$. □

We have not defined the linear transitivity of the group of all the collineations. The following theorem contains a first result. A more complete treatment will be given in Chapter 6.

THEOREM 8 *Let $P_1P_2P_3$ and $Q_1Q_2Q_3$ denote two triangles in a desarguesian plane. Then there are three axial affinities $\alpha_1, \alpha_2, \alpha_3$ such that*

$$P_k\alpha_1\alpha_2\alpha_3 = Q_k \quad \text{for } k = 1, 2, 3. \qquad (13)$$

Thus any given triangle can be mapped by a suitable collineation onto any other triangle.

PROOF Choose any line a_1 incident neither with P_1 nor with Q_1. Since the group $A(a_1)$ is linearly transitive by Theorem 7, there exists $\alpha_1 \in A(a_1)$ such that $P_1\alpha_1 = Q_1$. The image points $Q_1, P_2\alpha_1, P_3\alpha_1$ of the non-collinear points P_1, P_2, P_3 are again non-collinear.

Next choose a line a_2 through Q_1 which is incident neither with $P_2\alpha_1$ nor with Q_2. Then there is $\alpha_2 \in A(a_2)$ which maps $P_2\alpha_1$ onto Q_2. Since $Q_1 \, \mathrm{I} \, a_2$, we have

$$P_1\alpha_1\alpha_2 = Q_1\alpha_2 = Q_1, \quad \text{also } P_2\alpha_1\alpha_2 = Q_2.$$

The images $Q_1, Q_2, P_3\alpha_1\alpha_2$ of $Q_1, P_2\alpha_1, P_3\alpha_1$, respectively, are again non-collinear.

Finally put $a_3 = [Q_1, Q_2]$. Then $P_3\alpha_1\alpha_2$ and Q_3 are not incident with a_3, and there is an affinity α_3 in $A(a_3)$ which maps $P_3\alpha_1\alpha_2$ onto Q_3. As Q_1 and Q_2 are fixed points of α_3, $\alpha_1\alpha_2\alpha_3$ satisfies (13). □

EXERCISES

1 The group H of the homotheties of the plane $\mathfrak{A}$ is said to be linearly transitive if there is a homothety mapping P_1 onto Q_1 and P_2 onto Q_2 whenever $[P_1, P_2]$ and $[Q_1, Q_2]$ are parallel; cf. Chapter 2, Theorem 12. Show that an affine plane is desarguesian if and only if the group of its homotheties is linearly transitive.

2 The plane $\mathfrak{A}$ is desarguesian if and only if it is a translation plane and there is a point O such that $D(O)$ is linearly transitive.

3 Suppose that $D(O)$ is linearly transitive and that there is a collineation α such that $O \neq O\alpha$. Put $a = [O, O\alpha]$. Then $H(a)$ is linearly transitive. [Thus Theorem 2 can also be deduced from Theorem 3.]

4 Suppose $D(O)$ is linearly transitive and there are two collineations α, β such that O, $O\alpha$, $O\beta$ are not collinear. Then $\mathfrak{A}$ is desarguesian.

5 Suppose that $A(a)$ is linearly transitive and that there is a collineation α such that $a\alpha \neq a$. Then $\mathfrak{A}$ is desarguesian.

6 A *shear* is an axial affinity whose fixed lines are parallel to the axis. Let $\mathfrak{A}$ be desarguesian. Prove
(i) The product of two shears with axes in the parallel pencil Π is either a shear with an axis in Π or a translation in the direction of Π.
(ii) Every translation in the direction of Π is the product of two shears with axes in Π.
(iii) The union of the translation group $T(\Pi)$ with the set of the shears with axes in Π is a group.
(iv) This group is abelian.

7 The field F, the group F^+, and the operation $x \circ y$ are defined in Chapter 3, Exercise 9. The points of $\mathfrak{A}$ are again the ordered pairs (x, y) of elements of F. However, the lines are the sets

$\{(x, y) \mid x = c\}$ and $\{(x, y) \mid y = x \circ m + d\}$.

(i) Check that $\mathfrak{A}$ is an affine plane.
(ii) Show that $\mathfrak{A}$ is a translation plane.
(iii) Suppose r lies in F but not in the prime field of F; thus $r^p \neq r$. Prove that there is no dilatation with the centre O which maps the point $(1, 0)$ onto the point $(r, 0)$. Thus $\mathfrak{A}$ is a non-desarguesian translation plane.

EXERCISES

8* Show that (D) is equivalent to the following Statement $(\overline{D})$: Let l_1, l_2, l_3 denote three mutually distinct parallel lines. Let $P_1P_2P_3$ and $Q_1Q_2Q_3$ be two triangles such that P_i, $Q_i \mathbf{I} \, l_i$; $i = 1, 2, 3$. Then either corresponding sides of the two triangles are parallel or there is a line a such that any pair of corresponding sides of the triangles together with a belong to the same pencil.

5
Pappus planes

We have narrowed the class of all the affine planes first to that of the translation planes, then further to that of the desarguesian planes. In this chapter we introduce a subclass of the latter.

STATEMENT (P) ['Affine Theorem of Pappus']. *Suppose the lines l_1 and l_2 intersect, say at O. The points A_k, B_k, C_k are incident with l_k and distinct from O; $k = 1, 2$. Let*

$$[A_1, B_2] \parallel [B_1, A_2], \qquad [A_1, C_2] \parallel [C_1, A_2]. \tag{1}$$

Then

$$[B_1, C_2] \parallel [C_1, B_2]; \tag{2}$$

cf. Figure 5.1.

An affine plane is called a *Pappus plane* if the affine theorem (P) of Pappus is satisfied.

We note:

REMARK 1.1 *Statement* (P) *always holds if* A_1, B_1, C_1 *or* A_2, B_2, C_2 *are not mutually distinct*; cf. Chapter 3, Remark 2.2.

THEOREM 1 *Every affine Pappus plane is desarguesian.*

PROOF By Chapter 4, Theorem 5, it is sufficient to deduce Statement (S) from

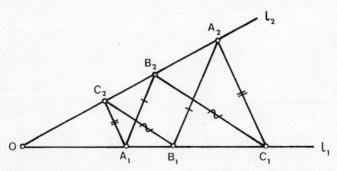

Figure 5.1

(P). This can be done by repeating the proof of Chapter 3, Theorem 4, merely replacing the parallel lines l_1 and l_2 by intersecting lines and Statement (p) by (P).

We can now reformulate the definition of Pappus planes.

STATEMENT (P') *Let l_1 and l_2 denote two distinct lines. Suppose the points A_k, B_k, C_k are incident with l_k but not with the other line; $k = 1, 2$. Then (1) implies (2).*

If $l_1 \nparallel l_2$, (P') becomes (P) while for $l_1 \parallel l_2$ we obtain (p).

COROLLARY 1.2 *An affine plane is a Pappus plane if and only if it satisfies Statement (P').*

PROOF Obviously, (P') implies that the given plane is a Pappus plane. Conversely, being desarguesian, a Pappus plane is a translation plane, by Chapter 4 Corollary 2.1. Hence by Chapter 3, Theorem 2, it satisfies not only (P) but also (p) and therefore (P'). □

We wish to describe Pappus planes in terms of their dilatation groups. Given a point O, (P)$_O$ will denote Statement (P) if the lines l_1 and l_2 are required to intersect at O.

THEOREM 2 *If the group $D(O)$ is linearly transitive and abelian, (P)$_O$ holds true. Conversely (P)$_O$ implies that $D(O)$ is abelian.*

PROOF
(i) Let $D(O)$ be linearly transitive and abelian. The lines l_k and the points A_k, B_k, C_k shall satisfy the assumptions of Statement (P)$_O$; $k = 1, 2$. In particular, we may assume (1).

Since $D(O)$ is linearly transitive, there are dilatations δ, δ' with centre O such that

$$B_1\delta = A_1, \qquad A_1\delta' = C_1.$$

Then by our assumptions, $A_2\delta = B_2$ and symmetrically $C_2\delta' = A_2$; cf. the proof of Chapter 4, Theorem 1. Hence $C_2\delta\delta' = C_2\delta'\delta = A_2\delta = B_2$. As $B_1\delta\delta' = A_1\delta' = C_1$ and $\delta\delta' \in D(O)$, we obtain

$$[B_1, C_2] \parallel [B_1, C_2]\delta\delta' = [B_1\delta\delta', C_2\delta\delta'] = [C_1, B_2].$$

This yields (2) and proves (P)$_O$.

(ii) Conversely, assume (P)$_O$. Let δ and δ' be dilatations with the centre O. We have to show

$$\delta\delta' = \delta'\delta. \tag{3}$$

Let OB_1C_2 be a triangle, $l_1 = [O, B_1]$, $l_2 = [O, C_2]$. Put

$$A_1 = B_1\delta, \qquad C_1 = A_1\delta', \qquad A_2 = C_2\delta', \qquad B_2 = A_2\delta.$$

Thus A_k, B_k, C_k are incident with l_k and distinct from O; $k = 1, 2$.

Since δ and δ' map a line onto a parallel line, we have

$[B_1, A_2] \parallel [B_1, A_2]\delta = [A_1, B_2]$ and $[A_1, C_2] \parallel [A_1, C_2]\delta' = [A_2, C_1]$.

Hence by Statement $(P)_O$,

$[B_1, C_2] \parallel [B_2, C_1]$. (4)

Since $\delta\delta'$ and $\delta'\delta$ map the line $[B_1, C_2]$ onto the parallel line through $B_1\delta\delta' = A_1\delta' = C_1$ and $C_2\delta'\delta = A_2\delta = B_2$, respectively, (4) implies

$[B_1, C_2]\delta\delta' = [B_1, C_2]\delta'\delta = [B_2, C_1]$.

In particular, $B_1\delta'\delta$ is the intersection of this line with l_1, i.e.

$B_1\delta'\delta = C_1 = B_1\delta\delta'$.

By Chapter 2, Theorem 22, there is only one dilatation with the centre O which maps B_1 onto C_1. Hence the last relation implies (3). □

Theorems 2 and 1 yield the group theoretic description of Pappus planes for which we have been looking:

COROLLARY 2.1 *An affine plane is a Pappus plane if and only if all the groups $D(O)$ are linearly transitive and abelian.*

We can now complete the diagrams of p. 42 to a diagram which contains the principal results of the last three sections:

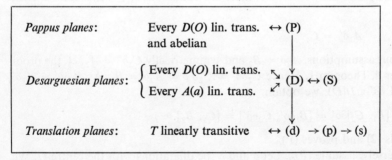

Diagram III

In Chapter 3, Exercise 9, finite planes were discussed which were not translation planes. In Chapter 4, Exercise 7, we had examples of finite non-desarguesian translation planes. In Chapter 6, we shall meet desarguesian planes which do not satisfy Statement (P). However, such planes are necessarily infinite.

APPENDIX

By Theorem 1, we have

THEOREM 1' *Statement* (P') *implies Statement* (D).

We wish to give a direct proof, cf. Figure 5.2.

Suppose the lines l_k and the points, O, P_k, Q_k satisfy the assumptions of Statement (D). Thus in particular

$$[P_0, P_1] \parallel [Q_0, Q_1] \quad \text{and} \quad [P_0, P_2] \parallel [Q_0, Q_2]. \tag{5}$$

We have to show that

$$[P_1, P_2] \parallel [Q_1, Q_2]. \tag{6}$$

By Chapter 4, Remark 1.1 we may assume that P_0, P_1, P_2 and Q_0, Q_1, Q_2 are not collinear and that $P_k \neq Q_k$ for $k = 0, 1, 2$.

If both $[P_1, P_2]$ and $[Q_1, Q_2]$ are parallel to l_0, (6) is trivial. Thus we may assume that, e.g.,

$$[P_1, P_2] \nparallel l_0. \tag{7}$$

The following construction will enable us to apply (P'). We shall not discuss the existence and non-incidence of the various points and lines; cf. the remark in the proof of Chapter 4, Theorem 2.

Let S denote the intersection of $[P_0, P_2]$ with the line through O parallel to $[P_1, P_2]$. Let R be the intersection of l_1 with the line through P_2 parallel to l_0. Then

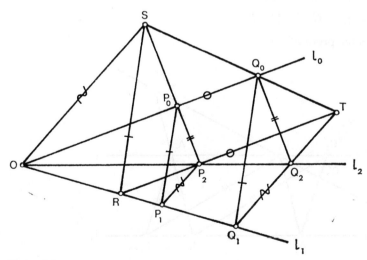

Figure 5.2

$[O, S] \parallel [P_1, P_2]$ and $[O, P_0] \parallel [R, P_2]$.

Hence by (P')

$[R, S] \parallel [P_1, P_0]$. (8)

The line $h = [S, Q_0]$ intersects $[P_2, R]$ at some point T. Then

$[S, P_2] \parallel [Q_0, Q_2]$ and $[Q_0, O] \parallel [T, P_2]$.

Thus (P') implies

$[O, S] \parallel [Q_2, T]$. (9)

If $Q_1 \, I \, h$, (8) implies

$h = [Q_0, Q_1] \parallel [P_0, P_1] \parallel [S, R]$.

Hence $h = [S, R]$. As this line intersects $[P_2, T]$ both at R and at T, this yields $R = T$. Since h intersects l_1 both at R and at Q_1, we obtain $R = T = Q_1$. Hence by (9),

$[P_1, P_2] \parallel [O, S] \parallel [Q_2, T] = [Q_2, Q_1]$.

This yields (6).

From now on we may assume $Q_1 \, \not{I} \, h$. Then

$[O, Q_0] \parallel [R, T]$ and $[R, S] \parallel [Q_1, Q_0]$,

Statement (P') now yields

$[O, S] \parallel [Q_1, T]$. (10)

By (9) and (10), $[T, Q_1] \parallel [T, Q_2]$. Hence

$[Q_1, Q_2] = [T, Q_1] \parallel [O, S] \parallel [P_1, P_2]$.

This completes the proof of (6). □

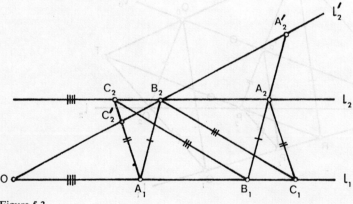

Figure 5.3

EXERCISES

1 Deduce (p) directly from (P); cf. Figure 5.3.

2 Show that every dilatation in a Pappus plane is the product of two suitable axial affinities.

3 Every translation plane contains Pappus subplanes of finite or countable order.

6
Co-ordinates in desarguesian planes

6.1
CO-ORDINATE PLANES

Let **V** be a two-dimensional right vector space over the skew field F and let O, A, B denote elements of **V**. By means of **V** we can construct a desarguesian plane. Our *points* will be the vectors of **V**. A *line* is a *point set*

$$l = A + \langle D \rangle = \{A + Dx \mid x \in F\}, \qquad D \neq O. \tag{1}$$

D corresponds to a 'direction vector' of l. Two lines are *equal* if they are identical point sets. Let L denote the set of the lines. The point P is said to be *incident* with the line l if $P \in l$. Thus two lines are *parallel* if, considered as point sets, they are either identical or disjoint. We have now defined a triplet

$$\mathfrak{A} = (\mathbf{V}, \mathbf{L}, \in).$$

In this plane, lines have been defined as point sets. Nevertheless, it was important that we did not do so in the first chapter. We shall later complete every affine plane to a projective plane, and there will be complete symmetry between points and lines in such a plane: If its points are reinterpreted as lines and its lines as points, another projective plane is obtained.

We prepare our discussion by the following remark.

LEMMA 1.1 *Let*

$$l = A + \langle D \rangle, \quad l' = A' + \langle D' \rangle.$$

Then

(i) $\langle D \rangle = \langle D' \rangle$ *implies* $l \parallel l'$;
(ii) *if* $\langle D \rangle \neq \langle D' \rangle$, *then l and l' have exactly one point in common.*
[Here $\langle D \rangle = \{Dx \mid x \in F\}$.]

PROOF
(i) Let $\langle D \rangle = \langle D' \rangle$; thus $l' = A' + \langle D \rangle$. Obviously, the lines l and l' are either disjoint or identical. Thus $l \parallel l'$.

6.1 CO-ORDINATE PLANES

(ii) Let $\langle D \rangle \neq \langle D' \rangle$. Then D and D' are linearly independent. Thus they form a base of **V**. The equation

$$A + Dx = A' + D'x' \tag{2}$$

in the unknowns x and x' is equivalent to

$$A - A' = D(-x) + D'x'.$$

Since $A - A'$ can be written as a linear combination of D and D' in one and only one way, the last equation has one and only one solution x, x'. The same applies to (2). Thus l and l' have exactly one point in common. In particular, they are not parallel. □

LEMMA 1.2 $\mathfrak{A} = (\mathbf{V}, \mathbf{L}, \in)$ *is an affine plane.*

PROOF To Axiom A1: Let P and Q be two distinct points. Then $Q - P \neq O$ and the line $P + \langle Q - P \rangle$ passes through both P and Q.

By Lemma 1.1, there is not more than one line incident with both P and Q.

To Axiom A2: Given the point P and the line $l = A + \langle D \rangle$, Lemma 1.1 implies that the line $P + \langle D \rangle$ is parallel to l and passes through P and that it is the only line with both of these properties.

To Axiom A3: If the vectors A and B are linearly independent, the three points O, A, B are not collinear. □

THEOREM 1 $\mathfrak{A} = (\mathbf{V}, \mathbf{L}, \in)$ *is a desarguesian affine plane.* We call it the co-ordinate plane over the skew field F.

We have to show that the group $D(C)$ is linearly transitive for any choice of C in **V**.

Suppose C, P, Q are collinear and mutually distinct. We have to construct a dilatation δ with the centre C such that $P\delta = Q$.

Since Q lies on the line $\{C + (P - C)x \mid x \in F\}$ through C and P and since $Q \neq C$, there exists $d \neq 0$ in F such that

$$Q = C + (P - C)d.$$

Define the maps δ and δ' of **V** into itself through

$$X\delta = C + (X - C)d, \quad Y\delta' = C + (Y - C)d^{-1}.$$

Then

$$X\delta = Y \Leftrightarrow Y - C = (X - C)d \Leftrightarrow X - C = (Y - C)d^{-1}$$

or

$$X\delta = Y \Leftrightarrow Y\delta' = X.$$

Thus $\delta: V \to V$ is bijective and has the inverse $\delta^{-1} = \delta'$.

If $X \neq X'$, then $[X, X'] = X + \langle X' - X \rangle$ and $[X\delta, X'\delta] = X\delta + \langle X'\delta - X\delta \rangle$.

Here

$$X'\delta - X\delta = (C+(X'-C)d) - (C+(X-C)d) = (X'-X)d.$$

Hence by (2), $\langle X'\delta - X\delta \rangle = \langle X'-X \rangle$ and by Lemma 1.1
$[X\delta, X'\delta] \parallel [X, X']$.

By Chapter 2, Theorem 8, our bijection δ can be extended to a homothety δ of $\mathfrak{A}$. As it has the fixed point C, it belongs to $D(C)$. Finally $P\delta = Q$. $\square$

THEOREM 2 $\mathfrak{A} = (\mathbf{V}, \mathbf{L}, \in)$ *is a Pappus plane if and only if F is a field.*

PROOF Let δ and δ' be two dilatations with the centre C; say,

$$X\delta = C+(X-C)d, \qquad X\delta' = C+(X-C)d'.$$

Then

$$X\delta\delta' = C+(X\delta-C)d' = C+(X-C)dd'$$

and

$$X\delta'\delta = C+(X-C)d'd.$$

Thus $\delta\delta' = \delta'\delta$ if and only if $dd' = d'd$. $\square$

6.2
CO-ORDINATES IN DESARGUESIAN PLANES

Our next goal is the theorem that every desarguesian plane is isomorphic to a plane $(\mathbf{V}, \mathbf{L}, \in)$ over some skew field F.

Let $\mathfrak{A} = (\mathbf{P}, \mathbf{L}, \mathbf{I})$ be a desarguesian affine plane. Choose two distinct points O and E of $\mathfrak{A}$; $l = [O, E]$. Put

$$F = \{P \mid P \mathbf{I} \, l\}$$

and let $A, B, C, \ldots$ denote elements of F.

Since $\mathfrak{A}$ is, in particular, a translation plane, there is to each A exactly one translation which maps O onto A. We denote it by τ_A. Thus

$$O\tau_A = A \qquad \text{for all } A \in F. \tag{3}$$

Similarly if $A \neq O$ there is one and only one dilatation $\delta = \delta_A$ with the centre O which maps E onto A:

$$E\delta_A = A \qquad \text{for all } A \in F \setminus \{O\}. \tag{4}$$

Define

$$A+B = O\tau_A\tau_B; \tag{5}$$

6.2 CO-ORDINATES IN DESARGUESIAN PLANES

$$A \cdot B = E\delta_A\delta_B \quad \text{if } A \neq O, B \neq O, \tag{6}$$

$$A \cdot O = O \cdot A = O. \tag{7}$$

By (5) and (3),

$$A + B = A\tau_B. \tag{8}$$

Furthermore, $O\tau_{A+B} = A + B = O\tau_A\tau_B$. Since $\tau_A\tau_B$ is a translation, Chapter 2, Theorem 16 implies

$$\tau_{A+B} = \tau_A\tau_B. \tag{9}$$

By (6), (7), and (4),

$$A \cdot B = A\delta_B \quad \text{if } B \neq O. \tag{10}$$

If A and B are distinct from O, (6) implies $A \cdot B \neq O$. By (4) and (6), $E\delta_{A \cdot B} = A \cdot B = E\delta_A\delta_B$. Hence by Chapter 2, Theorems 21 and 22,

$$\delta_{A \cdot B} = \delta_A\delta_B \quad \text{if } A \neq O, B \neq O. \tag{11}$$

LEMMA 3.1 *F is a skew field.*

PROOF
(i) By (9), the mapping of each τ_A onto the corresponding point A is an isomorphism of the abelian group of the translations in the direction of l onto F. Thus F is an abelian group under addition. For the convenience of the reader, we elaborate this argument.

By (9) and Chapter 2, Theorem 19,

$$\tau_{A+B} = \tau_A\tau_B = \tau_B\tau_A = \tau_{B+A}.$$

Hence by (3),

$$A + B = O\tau_{A+B} = O\tau_{B+A} = B + A.$$

Similarly by (9),

$$\tau_{(A+B)+C} = \tau_{A+B}\tau_C = \tau_A\tau_B\tau_C = \tau_A\tau_{B+C} = \tau_{A+(B+C)}$$

and thus

$$(A+B) + C = A + (B+C).$$

Since $\tau_O = \iota$, (9) implies

$$\tau_{A+O} = \tau_A\tau_O = \tau_A \quad \text{for all } A \in F.$$

Given A, put $B = O\tau_A^{-1}$. Then $\tau_B = \tau_A^{-1}$ and

$$\tau_{A+B} = \tau_A\tau_B = \tau_A\tau_A^{-1} = \iota = \tau_O.$$

(ii) The analogous arguments show that F satisfies the multiplicative laws of a skew field.

(iii) It remains to prove the distributive laws
$$(A+B)\cdot C = A\cdot C + B\cdot C \tag{12}$$
and
$$C\cdot(A+B) = C\cdot A + C\cdot B. \tag{13}$$

We may again assume that C is distinct from O.

Proof of (12). By Chapter 2, Theorem 17, $\delta_B^{-1}\tau_A\delta_B$ is a translation. By (3) and (10),
$$O\delta_B^{-1}\tau_A\delta_B = O\tau_A\delta_B = A\delta_B = A\cdot B = O\tau_{A\cdot B}.$$

Hence $\delta_B^{-1}\tau_A\delta_B = \tau_{A\cdot B}$ and
$$\tau_{(A+B)C} = \delta_C^{-1}\tau_{A+B}\delta_C = \delta_C^{-1}\tau_A\tau_B\delta_C = \delta_C^{-1}\tau_A\delta_C\,\delta_C^{-1}\tau_B\delta_C$$
$$= \tau_{A\cdot C}\tau_{B\cdot C} = \tau_{A\cdot C + B\cdot C}.$$

This yields (12).

For the *proof of* (13), we choose a line a through O distinct from l. By Chapter 4, Theorem 6, the group $A(a)$ of the affinities with the axis a is linearly transitive. Thus there is to every $A \in F$, $A \neq O$ one and only one $\alpha \in A(a)$ such that $E\alpha = A$. Write $\alpha = \alpha_A$. Thus
$$E\alpha_A = A \quad \text{for all } A \in F \setminus \{O\}. \tag{14}$$

By Chapter 2, Theorem 29, we have $\alpha_B\delta_A = \delta_A\alpha_B$. Hence
$$A\alpha_B = E\delta_A\alpha_B = E\alpha_B\delta_A = B\delta_A = B\cdot A.$$

This formula yields
$$\alpha_B^{-1}\tau_A\alpha_B = \tau_{B\cdot A},$$

since $\alpha_B^{-1}\tau_A\alpha_B$ is a translation which maps O onto
$$O\alpha_B^{-1}\tau_A\alpha_B = O\tau_A\alpha_B = A\alpha_B = B\cdot A = O\tau_{B\cdot A}.$$

We now obtain
$$\tau_{C\cdot(A+B)} = \alpha_C^{-1}\tau_{A+B}\alpha_C = (\alpha_C^{-1}\tau_A\alpha_C)(\alpha_C^{-1}\tau_B\alpha_C)$$
$$= \tau_{C\cdot A}\tau_{C\cdot B} = \tau_{C\cdot A + C\cdot B}.$$

This completes the proof of our lemma.

THEOREM 3 *Every desarguesian plane is isomorphic to a plane over a skew field.*

PROOF Given the desarguesian plane $A = (\mathbf{P}, \mathbf{L}, \mathrm{I})$, let
$$l = [O, E] \in \mathbf{L}, \quad F = \{P \mid P \,\mathrm{I}\, l\}.$$

6.2 CO-ORDINATES IN DESARGUESIAN PLANES

Define addition and multiplication in F by (5)–(7). Thus F becomes a skew field. The ordered pairs
$$\bar{A} = (A_1, A_2)$$
of elements of F form a two-dimensional right vector space $\bar{V}$ over F if we define
$$(A_1, A_2) + (B_1, B_2) = (A_1 + B_1, A_2 + B_2)$$
and
$$(A_1, A_2)C = (A_1 \cdot C, A_2 \cdot C).$$
By Theorem 2, $\bar{V}$ determines a desarguesian plane
$$\bar{\mathfrak{A}} = (\bar{V}, \bar{L}, \in)$$
where the elements of $\bar{L}$ are the point sets
$$\bar{A} + \langle \bar{B} \rangle = \{\bar{A} + \bar{B} \cdot C \mid C \in F\}.$$
We wish to construct a collineation of $\mathfrak{A}$ onto $\bar{\mathfrak{A}}$.

(i) Choose two lines a_1 and a_2 through O distinct from one another and from l. Given any point A in $\mathfrak{A}$, the lines through A parallel to a_2 and a_1 intersect l at two points A_1 and A_2, respectively. Then
$$A\varphi = (A_1, A_2) = \bar{A} \tag{15}$$
defines a map φ of $\mathbf{P}$ into $\bar{V}$. Obviously, φ is a bijection of $\mathbf{P}$ onto $\bar{V}$.

(ii) We next investigate how φ transforms the translations and dilatations of $\mathfrak{A}$; cf. Figure 6.1.

Let Π_s be a parallel pencil: $s \not\parallel l$. The projection ψ parallel to s onto l maps each point A onto the intersection $A\psi$ of l with the line of Π_s through A. Finally let τ^A denote the translation which maps A onto $A\psi$. Our definitions imply
$$O\tau_{A\psi} = A\psi = A\tau^A = O\tau_A\tau^A.$$
Hence
$$\tau_{A\psi} = \tau_A \tau^A$$
and
$$A\psi \tau_{B\psi} = A\tau^A \tau_B \tau^B = A\tau_B(\tau^A \tau^B).$$

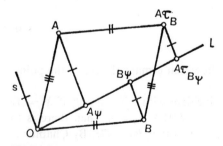

Figure 6.1

Since $A\psi$ and $B\psi$ lie on l, we have $(A\psi)\tau_{B\psi} \text{ I } l$. Also $\tau^A\tau^B \in T(\Pi_s)$. Thus $A\tau_B(\tau^A\tau^B)$ is the intersection of l with the line of Π_s through $A\tau_B$, i.e. it is the point $A\tau_B\psi$. This yields

$$(A\tau_B)\psi = A\psi\tau_{B\psi} = A\psi + B\psi \quad \text{for every } A, B. \tag{16}$$

Specializing $s = a_1$ or $s = a_2$, we have $A\psi = A_1$, and $A\psi = A_2$ respectively. The map $\varphi : \mathbf{P} \to \overline{\mathbf{V}}$ was defined by (15). Put

$$P\varphi = (P_1, P_2), \quad (P\tau_A)\varphi = (P_1', P_2').$$

If A is replaced by P and B by A, (16) yields $P_1' = P_1 + A_1$ and $P_2' = P_2 + A_2$. Thus

$$(P_1', P_2') = (P_1, P_2) + (A_1, A_2).$$

Hence

$$(P\tau_A)\varphi = P\varphi + A\varphi \quad \text{for all } P. \tag{17}$$

The discussion of the dilatations is similar. Let $\delta \in D(O)$, $E\delta = A$. Thus $\delta = \delta_A$. Put

$$P\varphi = (P_1, P_2), \quad (P\delta_A)\varphi = (P_1', P_2').$$

Since l is a trace of δ_A, we have $P_1\delta_A \text{ I } l$. Furthermore

$$a_2 \parallel [P, P_1] \parallel [P, P_1]\delta_A = [P\delta_A, P_1\delta_A].$$

Thus $P_1\delta_A$ is the intersection of l with the line through $P\delta_A$ parallel to a_2. Hence $P_1' = P_1\delta_A = P_1 \cdot A$; cf. (10). Symmetrically $P_2' = P_2 \cdot A$. Therefore

$$(P_1', P_2') = (P_1 \cdot A, P_2 \cdot A) = (P_1, P_2) \cdot A$$

or

$$(P\delta_A)\varphi = (P\varphi) \cdot A \quad \text{for all } P. \tag{18}$$

Put $Q = P\delta_A$; thus $P = Q\delta_A^{-1}$. Then (18) yields

$$Q\varphi = (Q\delta_A^{-1}\varphi) \cdot A$$

or

$$Q\delta_A^{-1}\varphi = Q\varphi \cdot A^{-1} \tag{19}$$

for all $A \text{ I } l$, $A \neq O$ and all Q.

(iii) The preceding remarks enable us to show that φ maps the set of the points on a line onto a line of $\overline{\mathfrak{A}} = (\overline{\mathbf{V}}, \overline{\mathbf{L}}, \in)$.

Given a line g and $A \text{ I } g$, the point X is incident with g if and only if $X = A\tau$ for some translation $\tau \in T(\Pi_g)$. Next, let $\tau_B \in T(\Pi_g)$, $\tau_B \neq \iota$. Then there is to every $\tau \in T(\Pi_g)$, $\tau \neq \iota$, a dilatation $\delta \in D(O)$ such that $B\delta = O\tau$. By Chapter 2, Theorems 17 and 16,

6.2 CO-ORDINATES IN DESARGUESIAN PLANES

$O\delta^{-1}\tau_B\delta = O\tau_B\delta = B\delta = O\tau$

implies

$\tau = \delta^{-1}\tau_B\delta.$

Putting $\delta = \delta_Y$, we obtain: Let $X \neq A$. Then $X \mathbf{I} g$ if and only if

$$X = A\delta_Y^{-1}\tau_B\delta_Y \qquad (20)$$

for some $Y \mathbf{I} l$, $Y \neq O$.

By (18), (17), and (19),

$(A\delta_Y^{-1}\tau_B\delta_Y)\varphi = (A\delta_Y^{-1}\tau_B)\varphi \cdot Y = (A\delta_Y^{-1}\varphi + B\varphi)Y = (A\varphi \cdot Y^{-1} + B\varphi)Y.$

Hence (20) is equivalent to

$X\varphi = A\varphi + B\varphi \cdot Y,$ i.e. to $X\varphi \in A\varphi + \langle B\varphi \rangle.$

Altogether we obtain

$X \mathbf{I} g \Leftrightarrow X\varphi \in A\varphi + \langle B\varphi \rangle.$

Thus φ preserves collinearity. By Chapter 2, Theorem 3, φ can be extended to a collineation of $\mathfrak{A}$ onto $\overline{\mathfrak{A}}$. Hence $\mathfrak{A}$ and $\overline{\mathfrak{A}}$ are isomorphic. □

COROLLARY 3.2 $\mathfrak{A}$ *is a Pappus plane if and only if F is a field.*

PROOF Since collineations of affine planes preserve incidence and parallelism, Statement (P) will hold true in $\mathfrak{A}$ if and only if it holds true in $\overline{\mathfrak{A}}$. By Theorem 2, $\overline{\mathfrak{A}}$ is a Pappus plane if and only if F is a field. □

Let $\mathfrak{A} = (\mathbf{P}, \mathbf{L}, \in)$ again denote an affine desarguesian plane. The skew field

$F = F(\mathfrak{A}; O, E)$

constructed in Lemma 3.1 depended on the choice of the points O and E. We next show:

THEOREM 4 *Let φ denote a collineation of the desarguesian plane $\mathfrak{A}$ onto the plane $\mathfrak{A}\varphi$. Then $\mathfrak{A}\varphi$ is desarguesian and*

$F(\mathfrak{A}; O, E) \simeq F(A\varphi; O\varphi, E\varphi).$

PROOF The first assertion is trivial. Let $l = [O, E]$. Then

$F(\mathfrak{A}; O, E) = \{A \mid A \mathbf{I} l\}$

has the null-element O and the one-element E. Similarly

$F(\mathfrak{A}\varphi; O\varphi, E\varphi) = \{A\varphi \mid A\varphi \mathbf{I} l\varphi\} = \{A\varphi \mid A \mathbf{I} l\}$

has the null-element $O\varphi$ and the one-element $E\varphi$. If $\Pi = \Pi_l$, then $\Pi\varphi = \Pi_{l\varphi}$

and φ induces isomorphisms of $T(\Pi)$ onto $T(\Pi\varphi)$ and of $D(O)$ onto $D(O\varphi)$; more precisely

$$\tau_{A\varphi} = \varphi^{-1}\tau_A\varphi \quad \text{for all } A \mathbin{\text{I}} l$$

and

$$\delta_{A\varphi} = \varphi^{-1}\delta_A\varphi \quad \text{for all } A \mathbin{\text{I}} l, A \neq O.$$

Hence

$$\tau_{(A+B)\varphi} = \varphi^{-1}\tau_{A+B}\varphi = \varphi^{-1}\tau_A\tau_B\varphi = (\varphi^{-1}\tau_A\varphi)(\varphi^{-1}\tau_B\varphi)$$
$$= \tau_{A\varphi}\tau_{B\varphi} = \tau_{A\varphi+B\varphi}.$$

Thus

$$(A+B)\varphi = A\varphi + B\varphi$$

for all A, B on l and similarly

$$(A \cdot B)\varphi = A\varphi \cdot B\varphi \tag{21}$$

first if $A \neq O$. Since $(O \cdot B)\varphi = O\varphi = O\varphi \cdot B\varphi$, (21) remains valid for every choice of A and B. □

COROLLARY 4.1 *Let $\mathfrak{A}$ denote a desarguesian plane. Let $O, E,$ and O', E' denote two pairs of points; $O \neq E, O' \neq E'$. Then*

$$F(\mathfrak{A}; O, E) \simeq F(\mathfrak{A}; O', E').$$

PROOF By Chapter 4, Theorem 8, there is a collineation φ of $\mathfrak{A}$ mapping O onto O' and E onto E'. Thus Theorem 4 yields our assertion. □

6.3*
THE FUNDAMENTAL THEOREM OF AFFINE GEOMETRY

Any affine desarguesian plane $\mathfrak{A}$ could be mapped isomorphically onto the co-ordinate plane $(\mathbf{V}, \mathbf{L}, \in)$ of some skew field F. This collineation induces an isomorphism of the group Γ of the collineations of $\mathfrak{A}$ onto the group of those of $(\mathbf{V}, \mathbf{L}, \in)$; cf. Chapter 2, Exercise 2. Thus we may assume in our study of Γ that $\mathfrak{A} = (\mathbf{V}, \mathbf{L}, \in)$.

We prepare our main theorem by two lemmas.

Choose any point O. Let τ_A denote the translation which maps O onto A.

LEMMA 5.1 *Every collineation of a desarguesian affine plane is the product of a collineation α_O with the fixed point O and a translation.*

PROOF Put $O_\alpha = C$ and $\alpha_O = \alpha\tau_C^{-1}$. Then $\alpha = \alpha_O\tau_C$ and $O\alpha_O = O\alpha\tau_C^{-1} = C\tau_C^{-1} = O$. □

6.3 THE FUNDAMENTAL THEOREM OF AFFINE GEOMETRY

LEMMA 5.2 *Let α be a collineation of a desarguesian affine plane; $O\alpha = O$. Then*

$$\tau_A{}^\alpha = \alpha^{-1}\tau_A\alpha = \tau_{A\alpha} \quad \text{for all } A. \tag{22}$$

This follows immediately from Chapter 2, Theorems 17 and 16 because

$$O\alpha^{-1}\tau_A\alpha = O\tau_A\alpha = A\alpha. \square$$

THEOREM 5 ['The Fundamental Theorem of Affine Geometry']. *The group Γ of all the collineations of the desarguesian affine plane $\mathfrak{A} = (\mathbf{V}, \mathbf{L}, \in)$ is identical with the group of the mappings α given by*

$$X\alpha = X\varphi + C \quad \text{for all } X \in \mathbf{V}. \tag{23}$$

Here C ranges through $\mathbf{V}$ and φ through the group $\Gamma(\mathbf{V})$ of the [semi-linear] automorphisms of the vector space $\mathbf{V}$.

We divide the proof into several steps.
(i) Every mapping α defined by (23) is a collineation of $\mathfrak{A}$: Certainly, α is a bijection of $\mathbf{V}$. Moreover, if σ is the automorphism of F belonging to φ, we have

$$(A+\langle B\rangle)\alpha = \{(A+Bx)\alpha \mid x \in F\} = \{(A+Bx)\varphi + C \mid x \in F\} \tag{24}$$
$$= \{A\varphi + C + B\varphi \cdot x^\sigma \mid x \in F\} = \{A\alpha + B\varphi \cdot y \mid y \in F\}$$
$$= A\alpha + \langle B\varphi\rangle.$$

Thus α is a collineation.
Obviously, we have
(ii) The collineations (23) form a subgroup of Γ. We denote it by Δ.
(iii) Δ contains all the translations of $\mathfrak{A}$: Given a translation $\tau = \tau_C$, we have

$$X\tau_C = X + C \quad \text{for all } X \in \mathbf{V}; \tag{25}$$

cf. (17). Thus τ_C is given by (23) with $\varphi = \iota$; hence $\tau_C \in \Delta$.
(iv) $D(O) \subset \Delta$: Let $\delta \in D(O)$. By (18), there is an $a \in F^* = F \setminus \{O\}$ such that

$$X\delta = X\delta_a = Xa \quad \text{for all } X \in \mathbf{V}. \tag{26}$$

This defines an automorphism (δ_a, ν) of the vector space $\mathbf{V}$, where ν denotes the inner automorphism $x \to a^{-1}xa$ of F induced by a. By (26), $\delta \in \Delta$.

From (25) and (26) we deduce

$$\tau_{A+B} = \tau_A \tau_B \quad \text{for all points } A, B, \tag{27}$$

$$\delta_{ab} = \delta_a \delta_b \quad \text{for all } a, b \text{ in } F^*. \tag{28}$$

(v) Let α be any collineation of $\mathfrak{A}$ with the fixed point O. We wish to show that the mapping $\alpha : \mathbf{V} \to \mathbf{V}$ is semilinear.
By (27), the mapping φ defined by

$$X\varphi = \tau_X \quad \text{for all } X \in \mathbf{V}$$

is an isomorphism of the additive group **V** onto the group T of the translations. By Lemma 5.2, the mapping ψ given by

$$\tau_X \psi = \tau_{X\alpha} \quad \text{for all } X \in \mathbf{V}$$

is an automorphism of T. Thus $\varphi\psi\varphi^{-1}$ is an automorphism of the additive group of **V**.

Since

$$X\varphi\psi\varphi^{-1} = \tau_X\psi\varphi^{-1} = \tau_{X\alpha}\varphi^{-1} = X\alpha \quad \text{for all } X \in \mathbf{V},$$

we have $\varphi\psi\varphi^{-1} = \alpha$ and

$$(X+Y)\alpha = X\alpha + Y\alpha \quad \text{for all } X, Y \text{ in } \mathbf{V}. \tag{29}$$

Similarly, define the mapping $\rho: F^* \to D(O)$ through

$$a\rho = \delta_a \quad \text{for all } a \in F^*$$

and the mapping η of $D(O)$ onto itself through

$$\delta^\eta = \alpha^{-1}\delta\alpha. \tag{30}$$

By (28), ρ is an isomorphism of the multiplicative group F^* onto $D(O)$ and η is an automorphism of F^*. Hence $\rho\eta\rho^{-1}$ is an automorphism of F^*. Put

$$a\mu = \begin{cases} 0 & \text{if } a = 0, \\ a\rho\eta\rho^{-1} & \text{if } a \neq 0. \end{cases} \tag{31}$$

Then by (30)

$$\alpha^{-1}\delta_a\alpha = \delta_{a\mu} \quad \text{for all } a \in F^*.$$

Hence

$$(Xa)\alpha = (X\delta_a)\alpha = (X\alpha)\delta_{a\mu} = X\alpha \cdot a\mu \quad \text{if } a \in F^*.$$

Obviously $(Xa)\alpha = X\alpha \cdot a\mu$ if $a = 0$. Thus

$$(Xa)\alpha = X\alpha \cdot a\mu \quad \text{for all } X \in \mathbf{V}, a \in F. \tag{32}$$

By (29) and (32), $\alpha: \mathbf{V} \to \mathbf{V}$ is semi-linear provided μ is an automorphism of F.

Since $\rho\eta\rho^{-1}$ is an automorphism of F^* and $0\mu = 0$, we have

$$(a \cdot b)\mu = (a\mu) \cdot (b\mu) \quad \text{for all } a, b \text{ in } F.$$

Furthermore, let E be any point $\neq O$. Then by (29) and (32)

$$(E\alpha)(a\mu + b\mu) = E\alpha \cdot a\mu + E\alpha \cdot b\mu = (Ea)\alpha + (Eb)\alpha$$
$$= (Ea + Eb)\alpha = (E(a+b))\alpha = E\alpha \cdot (a+b)\mu.$$

Hence $(a+b)\mu = a\mu + b\mu$ for all a, b in F. This proves (v). We restate the result:
(vi) Let α be any collineation of $\mathfrak{A}$ with the fixed point O. Define the auto-

6.3 THE FUNDAMENTAL THEOREM OF AFFINE GEOMETRY

morphism μ of F through (31). Then (α, μ) is an automorphism of $\mathbf{V}$ if we interpret α as a bijection of $\mathbf{V}$. Hence every collineation of $\mathfrak{A}$ with the fixed point O belongs to Δ; cf. (23).

(vii) *Proof of Theorem 5.* By (ii), we have $\Delta \subset \Gamma$. Conversely, let $\alpha \in \Gamma$. By Lemma 5.1, there are a collineation α_O of $\mathfrak{A}$ with the fixed point O and a translation τ such that $\alpha = \alpha_O \tau$. By (vi), $\alpha_O \in \Delta$; and by (iii), $\tau \in \Delta$. Thus $\alpha = \alpha_O \tau \in \Delta$ and $\Gamma \subset \Delta$. Hence $\Gamma = \Delta$. □

A product of axial affinities [of any affine plane] is called an *affinity*.

COROLLARY 5.1 *The group of all the affinities of a desarguesian affine plane $\mathfrak{A} = (\mathbf{V}, \mathbf{L}, \in)$ is identical with the group of those collineations (23) for which φ is a linear automorphism of the vector space $\mathbf{V}$.*

Our *proof* continues the preceding discussion.

Let Λ denote the group of those collineations given by (23) for which φ is a linear automorphism of $\mathbf{V}$.

(i) Λ contains the axial affinities and hence all the affinities: Let α be any axial affinity. By Theorem 5, there exists an automorphism (φ, μ) of $\mathbf{V}$ and a vector C such that (23) holds.

If A, B are two distinct points on the axis of α, that axis is the line $A + \langle B - A \rangle$. Hence for any $a \in F$

$$A + (B-A)a = (A + (B-A)a)\alpha = A\alpha + ((B-A)a)\varphi$$
$$= A\alpha + (B-A)\varphi \cdot a\mu = A\alpha + (B\varphi - A\varphi) \cdot a\mu$$
$$= A\alpha + (B\alpha - A\alpha) \cdot a\mu = A + (B-A) \cdot a\mu.$$

Thus $a = a\mu$ for every $a \in F$, $\mu = \iota$ and $\alpha \in \Lambda$.

In order to prove the converse, we first show

(ii) A collineation $\alpha \in \Lambda$ with three non-collinear fixed points P, Q, R is the identity.

To our α there exists a linear automorphism φ of $\mathbf{V}$ and a vector $C \in \mathbf{V}$ satisfying (23). Being linearly independent, the vectors $Q-P$, $R-P$ form a base of $\mathbf{V}$. We have

$$(Q-P)\varphi = Q\varphi - P\varphi = Q\alpha - P\alpha = Q-P$$

and similarly $(R-P)\varphi = R-P$. Since φ is linear, this implies $\varphi = \iota$. Furthermore $P = P\alpha = P\varphi + C = P + C$. Thus $C = O$ and $\alpha = \iota$.

(iii) Any collineation α in Λ is an affinity.

Choose any triangle PQR. By Chapter 4, Theorem 8, there exists an affinity β mapping PQR onto $P\alpha \, Q\alpha \, R\alpha$. Hence $\gamma = \alpha\beta^{-1}$ has the fixed points P, Q, R. By (i), $\gamma \in \Lambda$ and by (ii) $\gamma = \iota$. Thus $\alpha = \beta$ is an affinity. Combining (i) and (iii), we obtain Corollary 5.1. □

D

The preceding proof together with Chapter 4, Theorem 8, yields the following

COROLLARY 5.2 *Given two triangles $P_1P_2P_3$ and $Q_1Q_2Q_3$ in a desarguesian plane, there exists one and only one affinity mapping P_i onto Q_i, $i = 1, 2, 3$.*

EXERCISES

1 Let $\mathbb{H}$ denote the set of all the linear combinations of the matrices

$$e = \begin{pmatrix} 1 & 0 & 0 & 0 \\ 0 & 1 & 0 & 0 \\ 0 & 0 & 1 & 0 \\ 0 & 0 & 0 & 1 \end{pmatrix}, i = \begin{pmatrix} 0 & -1 & 0 & 0 \\ 1 & 0 & 0 & 0 \\ 0 & 0 & 0 & -1 \\ 0 & 0 & 1 & 0 \end{pmatrix}, j = \begin{pmatrix} 0 & 0 & 1 & 0 \\ 0 & 0 & 0 & -1 \\ -1 & 0 & 0 & 0 \\ 0 & 1 & 0 & 0 \end{pmatrix}, k = \begin{pmatrix} 0 & 0 & 0 & 1 \\ 0 & 0 & 1 & 0 \\ 0 & -1 & 0 & 0 \\ -1 & 0 & 0 & 0 \end{pmatrix}$$

with real coefficients.
(i) Show that $\mathbb{H}$ is a vector space over $\mathbb{R}$.
(ii) Verify

$$i^2 = j^2 = k^2 = -e, \quad ij = -ji = k, \quad jk = -kj = i, \quad ki = -ik = j,$$

and show that $\mathbb{H}$ is closed under multiplication.
(iii) Show that every element of $\mathbb{H}$ which is different from the null-matrix has an inverse and prove that $\mathbb{H}$ is a skew field. [Hint: if $\xi = te + xi + yj + zk$ where t, x, y, z are real numbers, put $\bar{\xi} = te - (xi + yj + zk)$ and compute $\xi\bar{\xi}$.]

Let $a \in F^* = F \setminus \{0\}$. Show that the dilatation

$$X \to X\delta = (X - C)a + C$$

belongs to Λ if and only if a lies in the centre of the multiplicative group F^*.

3 (i) If the affinity α in a desarguesian plane has two distinct fixed points, the line through them is an axis. Thus a non-axial affinity has not more than one fixed point.
(ii) If α has a fixed point, it is the product of two axial affinities. Is the converse true?

4 Show that every central affinity in a desarguesian plane is a dilatation which is the product of two axial affinities; cf. Exercise 2.

5 Prove that every affinity in a desarguesian plane is the product of an axial affinity with two shears; cf. Exercise 4 and Chapter 4, Exercise 6.

6 Let α denote an affinity in a co-ordinate plane $(\mathbf{V}, \mathbf{L}, \in)$ with an axis through O; $\alpha \neq \iota$. Show that there is a base $\{A, B\}$ of $\mathbf{V}$ such that

$$A\alpha = A, \quad B\alpha = \begin{cases} Bd & \text{if } \alpha \text{ is not a shear } (d \in F), \\ A + B & \text{if } \alpha \text{ is a shear.} \end{cases}$$

7
The projective closure of an affine plane

7.1
MOTIVATION

In any affine plane $\mathfrak{A} = (\mathbf{P}, \mathbf{L}, \mathtt{I})$ we have to distinguish between pairs of parallel and pairs of intersecting lines. We wish to imbed $\mathfrak{A}$ into a system

$$\mathfrak{A}^* = (\mathbf{P}^*, \mathbf{L}^*, \mathtt{I}^*)$$

in which this distinction is eliminated. Thus any two distinct lines of $\mathfrak{A}^*$ shall always have exactly one point in common. To this end we introduce new points and a single additional line calling each parallel pencil an *improper point* and the set of the improper points the *improper line*. The symbols, $P_\omega, Q_\omega, \ldots$ will denote improper points; the improper line is denoted by l_ω. Since we wish to keep distinguishing between a line and the set of its points, we introduce a redundant symbol $\mathbf{P}_\omega$ for the set of all the points of l_ω. We now define

$$\begin{cases} \mathbf{P}^* = \mathbf{P} \cup \mathbf{P}_\omega, \\ \mathbf{L}^* = \mathbf{L} \cup \{l_\omega\}, \\ \mathtt{I}^* = \mathtt{I} \cup \{(P_\omega, l_\omega) \mid P_\omega \in \mathbf{P}_\omega\} \cup \{(P_\omega, l) \mid l \in \Pi, \text{ where } \Pi = P_\omega\}. \end{cases} \qquad (1)$$

Extending our terminology we say that P is *incident* (in $\mathfrak{A}^*$) with l if $P \mathtt{I}^* l$. By (1), this usage will not lead to contradictions.

By (1), the improper line l_ω is incident precisely with the improper points. An improper point P_ω and a line of $\mathbf{L}$ are incident if and only if l belongs to the corresponding parallel pencil.* By this definition, every *proper* line, i.e. every line in $\mathbf{L}$ is incident with one and only one improper point; and any two proper lines are parallel in $\mathfrak{A}$ if and only if they are incident in $\mathfrak{A}$ with the same improper point.

We next collect some of the basic properties of $\mathfrak{A}^*$. Since they are to be used afterwards as axioms for a new kind of planes, they will be stated in as weak a form as possible.

---

* Note that an improper point P_ω is the set of all the lines of a parallel pencil, i.e. of all the *proper* lines through P_ω. It is not the set of *all* the lines through P_ω in the projective closure of the plane; cf. p. 87.

THEOREM A

(i) *Given any two points of* **P***, *there always exists a line incident with both of them.*
(ii) *Given any two lines of* **L***, *there is always a point incident with both of them.*

PROOF We may assume that the given points P and Q and the given lines l and h are mutually distinct.
(i) If P, Q are both proper, our statement follows from Axiom A1. If P and Q are both improper, l_ω is incident with both; cf. the above definition. Finally assume that $P = P_\omega$ is improper while Q is proper. Let l denote the line of the parallel pencil P_ω through Q; cf. Axiom A2. Then by our definition, $l\,\text{I}^*\,Q, P_\omega$.
(ii) Suppose both l and h are proper. If $l \not\parallel h$, $P = [h, l]$ has the required property. If $l \parallel h$, there is an improper point P_ω such that $P_\omega\,\text{I}^*\,l, h$. Finally, let $l = l_\omega$ while h is proper. Then there exists an improper point P_ω such that $P_\omega\,\text{I}^*\,h, l_\omega$. □

We next verify

THEOREM B *If both of the points P and Q are incident with both lines l and h, then $P = Q$ or $l = h$.*

PROOF Let $l \neq h$. If l and h are both proper, $l \not\parallel h$, then these lines belong to different parallel pencils. Thus they have exactly one proper and no improper point in common.

If $l \parallel h$, say $\{l, h\} \subset \Pi$, they have no proper point in common but exactly one improper point, viz. $P_\omega = \Pi$.

Finally assume that $l = l_\omega$ while h is proper. If $\Pi = \Pi_h$, l and h have the point $P_\omega = \Pi$ but no other point in common. □

THEOREM C *There are four points in* **P*** *such that no three of them are collinear.*

PROOF By Axiom A3 there are three non-collinear points A, B, C in **P**. The lines through B parallel to $[A, C]$ and through C parallel to $[A, B]$ intersect at a point $D \in \mathbf{P}$. Thus

$$[A, B] \parallel [C, D], \quad [A, B] \neq [C, D].$$

This readily implies that no three of these points are collinear in $\mathfrak{A}$. Hence they are not collinear in $\mathfrak{A}^*$ either. □

7.2
THE AXIOMS OF A PROJECTIVE PLANE

We can now define a *projective plane* starting with two disjoint sets **P** of *points* and **L** of *lines* and an *incidence relation* $\text{I} \subset \mathbf{P} \times \mathbf{L}$. Capital and small letters denote points and lines, respectively. If $(P, l) \in \text{I}$, we write again $P\,\text{I}\,l$ or $l\,\text{I}\,P$.

7.2 THE AXIOMS OF A PROJECTIVE PLANE

The triplet

$$\mathfrak{P} = (\mathbf{P}, \mathbf{L}, \mathbf{I})$$

is called a *projective plane* if it satisfies the following three conditions.

AXIOM P1
(i) *To any pair of points there is a line incident with both.*
(ii) *To any pair of lines there is a point incident with both.*

AXIOM P2 *If both of the points P, Q are incident with both of the lines l, h, then $P = Q$ or $l = h$.*

AXIOM P3 *There are four points no three of which are collinear.*

We may write Axiom P2 as follows: $P, Q \mathbf{I} \, l, h$ implies $P = Q$ or $l = h$ [or both] *for any two points P, Q and lines l, h.*

If $P \neq Q$, the line incident with P and Q is uniquely determined; cf. Axiom P2. It will again be denoted by $[P, Q]$. Similarly, $[l, h]$ will denote the point incident with the distinct lines l and h.

Given any affine plane $\mathfrak{A} = (\mathbf{P}, \mathbf{L}, \mathbf{I})$, the triplet $\mathfrak{A}^* = (\mathbf{P}^*, \mathbf{L}^*, \mathbf{I}^*)$ as defined on p. 85 is called the *projective closure* of $\mathfrak{A}$. Theorems A–C can be restated as follows:

THEOREM 1 *The projective closure of an affine plane is a projective plane.*

Conversely, we can construct an affine plane by deleting any line and its points from a projective plane: Let $l = l_\omega$ be any line in the projective plane $\mathfrak{P} = (\mathbf{P}, \mathbf{L}, \mathbf{I})$. Put

$$\mathbf{P}_l = \mathbf{P} \setminus \{P \mid P \mathbf{I} \, l\},$$

$$\mathbf{L}_l = \mathbf{L} \setminus \{l\}.$$

$\mathbf{I}_l$ shall denote the restriction of the incidence relation in $\mathfrak{P}$ to the point set $\mathbf{P}_l$ and the set of lines $\mathbf{L}_l$; more precisely, $\mathbf{I}_l$ will be the intersection of $\mathbf{I}$ with $\mathbf{P}_l \times \mathbf{L}_l$.

THEOREM 2 $\mathbf{P}_l = (\mathbf{P}_l, \mathbf{L}_l, \mathbf{I}_l)$ *is an affine plane.*

PROOF To axiom A1. If P and Q are any two distinct points of $\mathbf{P}_l$, Axiom P1(i) implies that there is a line $h \in \mathbf{L}$ incident with both. By Axiom P2, h is unique. Since $P \not{I} \, l$, we have $h \neq l$ and thus $h \in \mathbf{L}_l$.

To Axiom A2. Let $h \in \mathbf{L}_l, P \in \mathbf{P}_l, P \not{I}_l h$. Thus $h \neq l$ and $P \not{I} h$. Put $Q = [h, l]$. Any line g through P is distinct from l. The point $[g, h]$ does not exist in $\mathbf{P}_l$ if and only if it is incident with l, i.e. if and only if it is equal to Q. Thus there is one and only one line through P parallel in $\mathfrak{P}_l$ to h, viz. the line $g = [P, Q]$.

To Axiom A3. By Axiom P3 there are four points A, B, C, D, no three of which are collinear. If, at most, one of them is incident with l, the remaining three satisfy Axiom A3. Since more than two of these points cannot be incident with l, there remains only the case that exactly two of them are, say, the points C and D.

By Axioms P1(ii) and P2, the lines $[A, D]$ and $[B, C]$ intersect at some point E. It must be distinct from one of the points A and B, e.g. from A. By our assumptions

$$[A, E] = [A, D] \neq [A, B].$$

By P2, $[A, E]$ and $[A, B]$ intersect at A and nowhere else. Hence $E \not{I} [A, B]$. Symmetrically, $E \not{I} [C, D] = l$. Hence $E \in \mathbf{P}_l$ and the triplet A, B, E is non-collinear in $\mathbf{P}_l$. □

While an affine plane can be completed in only one way to its projective closure, Theorem 2 shows that any projective plane is the projective closure (up to isomorphisms) of a great many affine planes; cf. Theorem 3.

If $\mathfrak{A} = (\mathbf{P}, \mathbf{L}, \mathbf{I})$ is any affine plane, its projective closure is uniquely determined. We denoted it by $\mathfrak{A}^* = (\mathbf{P}^*, \mathbf{L}^*, \mathbf{I}^*)$. Conversely, given a projective plane $\mathfrak{P} = (\mathbf{P}, \mathbf{L}, \mathbf{I})$ and any line l in $\mathfrak{P}$, if we designate l as the improper line, we obtained an affine plane denoted by $\mathfrak{P}_l$.

THEOREM 3 *Let $\mathfrak{A}$ be any affine plane, $l = l_\omega$ its improper line and $\mathfrak{A}^*$ the projective closure of $\mathfrak{A}$. Then*

$$(\mathfrak{A}^*)_l = \mathfrak{A}.$$

Conversely, let $\mathfrak{P}$ be any projective plane and l be any line in $\mathfrak{P}$. Then

$$\mathfrak{P}_l^* \simeq \mathfrak{P}.$$

PROOF See Exercise 2.

EXAMPLE 3.1 Let

$$\mathbf{P} = \{P, Q_1, \ldots, Q_n\}, \qquad \mathbf{L} = \{l, h_1, \ldots, h_n\}.$$

Define incidence through

$$P \mathbf{I} h_k, \quad Q_k \mathbf{I} h_k, \quad Q_k \mathbf{I} l; \quad k = 1, \ldots, n.$$

The triplet $(\mathbf{P}, \mathbf{L}, \mathbf{I})$ satisfies the axioms P1 and P2. However P3 does not hold true. If a line is omitted, we do not obtain an affine plane; cf. Exercise 3.

7.3
THE DUALITY PRINCIPLE

If the sets $\mathbf{P}$ of the points and $\mathbf{L}$ of the lines are interchanged, the axioms P1 and P2 remain unchanged. We examine Axiom P3. Three points were called collinear

7.3 THE DUALITY PRINCIPLE

if there is a line with which all of them are incident. Correspondingly, we call three lines *co-punctal* if there is a point with which all three of them are incident. Then Axiom P3 corresponds to the following condition:

AXIOM P3' *There are four lines, no three of which are co-punctal.*

THEOREM 4 *The sets of axioms* P1, P2, P3 *and* P1, P2, P3' *are equivalent.*

We show

P1 ∧ P2 ∧ P3 ⇒ P3'. (2)	P1 ∧ P2 ∧ P3' ⇒ P3. (2')
Choose the four points P_1, P_2, P_3, P_4 according to P3. Then the four lines	Choose the four lines l_1, l_2, l_3, l_4 according to P3'. Then the four points
$[P_1, P_2]$, $[P_2, P_3]$, $[P_3, P_4]$, $[P_4, P_1]$ (3)	$[l_1, l_2]$, $[l_2, l_3]$, $[l_3, l_4]$, $[l_4, l_1]$ (3')
are mutually distinct. By P2,	are mutually distinct. By P2,
$P_2 = [[P_1, P_2], [P_2, P_3]]$.	$l_2 = [[l_1, l_2], [l_2, l_3]]$.
Since P_2, P_3, P_4 are not collinear, P_2 is not incident with $[P_3, P_4]$. Hence the three lines	Since l_2, l_3, l_4 are not co-punctal, l_2 is not incident with $[l_3, l_4]$. Hence the three points
$[P_1, P_2]$, $[P_2, P_3]$, $[P_3, P_4]$	$[l_1, l_2]$, $[l_2, l_3]$, $[l_3, l_4]$
are not co-punctal. Permuting the indices 1, 2, 3, 4 cyclically, we obtain more generally that no three of the four lines (3) are co-punctal. □	are not collinear. Permuting the indices 1, 2, 3, 4 cyclically, we obtain more generally that no three of the four points (3') are collinear. □

By Theorem 4, any projective plane satisfies the axioms P1, P2, P3, and P3'. Conversely, any triplet (**P**, **L**, **I**) satisfying these axioms is, of course, a projective plane. This set of axioms is transformed into itself when **P** and **L** are interchanged.

Given a statement of plane projective geometry, we obtain the *dual* statement by interchanging the expressions 'point' and 'line.' Thus this concept of duality is symmetric: the dual of the dual statement is the original one. Obviously, a statement may be *self-dual*. We can also form the dual of a collection of statements.

EXAMPLES
(i) Axioms P3 and P3' are dual to one another.
(ii) Axiom P2 is self-dual.
(ii) The assertion (2) is dual to (2'). Thus Theorem 4 is self-dual.

(iv) The above proofs of (2) and (2') are dual.

We now state the

Duality Principle of (plane) **Projective Geometry:** *Suppose a theorem of plane projective geometry can be formulated using exclusively the concepts of points, lines, and incidence, and it can be proved using only the axioms of this geometry. Then the dual theorem also holds true.*

We obtain the proof of the dual theorem by interchanging the terms 'point' and 'line' and the Axioms P3 and P3' in the original proof. As a rule, other concepts and previously proved theorems will appear in such a proof. They will have to be replaced by their duals.

The duality principle can be extended by adding other assumptions to our axioms, provided only that their duals are either assumed too or can be deduced from them. Such assumptions will be similar to the statements (D) and (P) of affine geometry.

The duality principle is *meta-mathematical* ('beyond mathematics') in the following sense: it does not deal with the objects of a mathematical theory. Its object is rather a theory itself, viz. plane projective geometry.

Warning: The Duality Principle does *not* assert the existence of a pair of bijections mapping the sets of the points and lines of a projective plane onto the sets of the lines and points, respectively, *of the same plane* and preserving incidence.

The textbooks of projective geometry show that such *dualities* exist in projective pappus planes; cf. Chapter 8.

We next apply the duality principle to translate two results of Chapter 1 into the projective context. Let $\mathfrak{P} = (\mathbf{P}, \mathbf{L}, \mathrm{I})$ denote a projective plane.

THEOREM 5 *Any point in $\mathfrak{P}$ is incident with at least three lines.*

Let $P \in \mathbf{P}$. Let l denote any line not incident with P; cf. Axiom P3. Designating l as the improper line, we obtain an affine plane. By Chapter 1, Theorem 7(iii), P is incident with not less than three lines of that plane. □

The dual of Theorem 5 is

THEOREM 5' *Any line in $\mathfrak{P}$ is incident with at least three points.*

We can deduce Theorem 5' from Theorem 5 by applying the duality principle. Alternatively, Theorem 5' can readily be derived from Chapter 1, Theorem 7(ii).

7.4
COLLINEATIONS

Let $\mathfrak{P} = (\mathbf{P}, \mathbf{L}, \mathrm{I})$ and $\mathfrak{P}' = (\mathbf{P}', \mathbf{L}', \mathrm{I}')$ denote two projective planes. A pair of bijections π of $\mathbf{P}$ onto $\mathbf{P}'$ and of $\mathbf{L}$ onto $\mathbf{L}'$ is called a *collineation* or an

7.4 COLLINEATIONS

isomorphism if it preserves incidence, i.e. if $P\,\mathrm{I}\,l$ always implies $(P\pi)\,\mathrm{I}'\,(l\pi)$. A trivial example is the identity $\iota = \iota_{\mathbf{P}}$ defined by

$$P\iota = P, \quad l\iota = l \qquad \text{for all } P \in \mathbf{P}, l \in \mathbf{L}.$$

We state without proof.

THEOREM 6 *Every collineation π satisfies*

$$[P, Q]\pi = [P\pi, Q\pi] \qquad \text{for all } P \neq Q, \tag{4}$$

$$[l, h]\pi = [l\pi, h\pi] \qquad \text{for all } l \neq h. \tag{5}$$

LEMMA 7.1 *The inverse of a collineation is a collineation. Thus for any collineation π* $P\,\mathrm{I}\,l \Leftrightarrow (P\pi)\,\mathrm{I}'\,(l\pi)$.

THEOREM 7 *The collineations of a projective plane form a group.*

Let π be a collineation of $\mathfrak{P}$. If $P \neq P\pi$, we call $[P, P\pi]$ a *trace* of π. If $l \neq l\pi$, the point $[l, l\pi]$ may be called a *dual trace* of π. The point P is a *fixed point* of π if $P\pi = P$; if $l\pi = l, l$ is a *fixed line* of π. If all the points [lines] incident with a given line [point] are fixed, the latter is called an *axis* [a *centre*] of π. Obviously any axis is a fixed line and any centre is a fixed point. Conversely, every fixed line is either an axis or a trace and every fixed point is either a centre or a dual trace. An *axial* [*central*] collineation has an axis [a centre].

THEOREM 8 *If the collineation π is axial, every trace of π is fixed.*

PROOF Let a denote an axis of π, and let $[P, P\pi]$ be a trace. Thus $P \neq P\pi$ and $P, P\pi \not{\mathrm{I}}\, a$; in particular, $[P, P\pi] \neq a$. Let a and $[P, P\pi]$ intersect at Q. Then $Q\pi = Q$ and

$$[P, P\pi]\pi = [P, Q]\pi = [P\pi, Q\pi] = [P\pi, Q] = [P\pi, P].\;\square$$

THEOREM 9 *Let π be any axial collineation. Then any fixed point outside the axis is a centre. If $\pi \neq \iota$, there is at most one such point.*

PROOF Let a be an axis of π. If $O\pi = O$ and $O \not{\mathrm{I}}\, a$, any line l through O intersects a at a fixed point $\neq O$. Hence $l\pi = \pi$ and O is a centre of π.

Suppose π has the distinct fixed points $O, O' \not{\mathrm{I}}\, a$. Then O, O' are centres. If $P \not{\mathrm{I}}\, [O, O']$, the point P is the intersection of the distinct fixed lines $[O, P]$ and $[O', P]$; hence $P\pi = P$. If $P\,\mathrm{I}\,[O, O']$, we replace O' by any point O''; $\not{\mathrm{I}}\,[O, O']$, obtaining again $P\pi = P$. Thus $P\pi = P$ for every point P and $\pi = \iota$. $\square$

We can now prove

THEOREM 10 *A collineation is axial if and only if it is central.*

PROOF By the duality principle, it is sufficient to prove that an axial collineation π has a centre. We may assume $\pi \neq \iota$. Because of Theorem 9, we may also assume that π has no fixed point outside its axis a.

Choose any point $P_0 \not{I} a$. Then $P_0 \neq P_0\pi$ and the line $l_0 = [P_0, P_0\pi] \neq a$ is fixed by Theorem 8. The point $O = [a, l_0]$ is a centre of π: let l be any line through O. If $l = a$ or $l = l_0$ we have $l\pi = l$. Let $l \neq a, l_0$ and let P denote any point on l; $P \neq O$. By Theorem 8, the line $[P, P\pi]$ is fixed. Hence its intersection with l_0 is fixed by Theorem 6 and therefore lies on a. Thus l_0 and $[P, P\pi]$ intersect at $[a, l_0] = O$. Hence $l = [P, P\pi]$, and l is fixed. □

Because of Theorem 10, there is no distinction between axial and central collineations. We therefore define: a collineation π of a projective plane is *perspective* if it has an axis a and a centre O. If $O \, I \, a$, it is an *elation*; for $O \not{I} a$ it is a *homology*.

Let π be a perspective collineation with the axis a and the centre O. Then obviously

$A, A\pi, O$ are collinear for any point A; (6)

$l, l\pi, a$ are co-punctal for any line l. (7)

These relations motivate the name 'perspective collineations': Choose any three non-collinear points A_1, A_2, A_3 distinct from O and outside a. Then the triangles $A_1A_2A_3$ and $(A_1\pi)(A_2\pi)(A_3\pi)$ are perspective from the centre O and the axis a; cf. Chapter 8.

Since a perspective collineation cannot have more than one axis and centre, Theorem 9 implies:

THEOREM 11 *Let π denote any perspective collineation; $\pi \neq \iota$. Then the fixed points and lines of π are the centre and the axis of π, the points incident with the axis, and the lines incident with the centre.*

We define the following sets of perspective collineations π:

$P(O, a) = \{\pi \mid \pi$ has the centre O and the axis $a\}$,

$P(O) = \{\pi \mid \pi$ has the centre $O\}$,

$P(a) = \{\pi \mid \pi$ has the axis $a\}$,

$P(a, b) = \{\pi \mid$ the centre of π is on a; π has the axis $b\}$,

$P(A, B) = \{\pi \mid \pi$ has the centre A; the axis is through $B\}$.

By Theorem 11.

$P(O, a) \cap P(O', a') = \{\iota\}$ if $O \neq O'$ or $a \neq a'$. (8)

THEOREM 12 *The sets $P(O, a), P(O), P(a), P(a, b), P(A, B)$ are groups. Through*

$$\pi \mapsto \pi^\alpha = \alpha^{-1}\pi\alpha,$$

any collineation α defines isomorphisms of $P(O, a), P(O), P(a), P(a, b), P(A, B)$

7.4 COLLINEATIONS

onto $P(O\alpha, a\alpha)$, $P(O\alpha)$, $P(a\alpha)$, $P(a\alpha, b\alpha)$, $P(A\alpha, B\alpha)$ respectively. In particular, $P(O, a)$ is a normal subgroup of $P(O)$ and of $P(a)$ if $O \mathrel{\text{I}} a$.

PROOF The first assertion is obvious. Next assume, e.g., that $\pi \in P(O, a)$. Then

$$l \mathrel{\text{I}} O\alpha \Rightarrow l\alpha^{-1} \mathrel{\text{I}} O \Rightarrow l\alpha^{-1}\pi = l\alpha^{-1} \Rightarrow l\alpha^{-1}\pi\alpha = l\alpha^{-1}\alpha = l$$

and

$$P \mathrel{\text{I}} a\alpha \Rightarrow P\alpha^{-1} \mathrel{\text{I}} a \Rightarrow P\alpha^{-1}\pi = P\alpha^{-1} \Rightarrow P\alpha^{-1}\pi\alpha = P.$$

Thus $\alpha^{-1}\pi\alpha \in P(O\alpha, a\alpha)$ and therefore

$$\alpha^{-1}P(O, a)\alpha \subset P(O\alpha, a\alpha).$$

Replacing $P(O, a)$ and α by $P(O\alpha, a\alpha)$ and α^{-1}, respectively, we obtain

$$\alpha P(O\alpha, a\alpha)\alpha^{-1} \subset P(O, a) \text{ or } P(O\alpha, a\alpha) \subset \alpha^{-1}P(O, a)\alpha.$$

Hence

$$\alpha^{-1}P(O, a)\alpha = P(O\alpha, a\alpha).$$

If $O \mathrel{\text{I}} a$ and $\alpha \in P(O)$ or $\alpha \in P(a)$, then $O\alpha = O$ and $a\alpha = a$. Thus $\alpha^{-1}P(O, a)\alpha = P(O, a)$. □

The reader will readily verify

THEOREM 13 *Suppose the points A, B are distinct from O and non-incident with a. Then there is not more than one perspective collineation in $P(O, a)$ which maps A onto B. More generally: let $A, B \mathrel{\rlap{\text{I}}{/}} a, b$. Then there is not more than one collineation in $P(a, b)$ which maps A onto B. Dually, there is not more than one collineation in $P(A, B)$ which maps a onto b.*

We call these groups *linearly transitive* if these collineations actually exist, whenever possible. In detail:

$P(O, a)$ is linearly transitive if to every pair of points A, B non-incident with a such that O, A, B are collinear and mutually distinct, there exists a collineation in $P(O, a)$ which maps A onto B.

$P(a, b)$ is linearly transitive if to every pair of points $A, B \mathrel{\rlap{\text{I}}{/}} a, b$ there is a collineation in $P(a, b)$ which maps A onto B.

Finally, $P(A, B)$ is linearly transitive if to every pair of lines $a, b \mathrel{\rlap{\text{I}}{/}} A, B$ there is a collineation in $P(A, B)$ which maps a onto b.

REMARK Usually, the *plane* is called (O, a)-*transitive* if $P(O, a)$ is linearly transitive. It is (a, b)-*transitive* if $P(a, b)$ is linearly transitive; etc.

The following theorem readily yields necessary conditions for the product of two perspective collineations to be again a perspective collineation.

THEOREM 14 *Let $\pi_\lambda \in P(O_\lambda, a_\lambda)$; $\pi_\lambda \neq \iota$ for $\lambda = 1, 2, 3$; $\pi_1\pi_2\pi_3 = \iota$.*

Then one of the following conditions is satisfied:

$$O_1 = O_2 = O_3 \quad \text{and} \quad a_1 = a_2 = a_3. \tag{9}$$

$$\begin{cases} O_1 = O_2 = O_3; \\ \text{the } a_\lambda \text{ are co-punctal and mutually distinct.} \end{cases} \tag{10}$$

$$\begin{cases} a_1 = a_2 = a_3; \\ \text{the } O_\lambda \text{ are collinear and mutually distinct.} \end{cases} \tag{11}$$

$O_1 O_2 O_3$ is a triangle such that

$$a_1 = [O_2, O_3], \quad a_2 = [O_3, O_1], \quad a_3 = [O_1, O_2]. \tag{12}$$

PROOF If, e.g., $O_1 = O_2$, this point is also the centre of $\pi_3 = \pi_2^{-1}\pi_1^{-1}$. If $a_1 = a_2$, this line is also the axis of π_3. From now on we may exclude (9).

Next assume $O_1 = O_2 = O_3 = O$. Thus the lines a_λ are mutually distinct. Suppose they are not co-punctal. Then two of their three intersections are distinct from O. Let $Q = [a_1, a_2] \neq O$. Since Q is a fixed point of π_1 and π_2, Q is also one of π_3. Since $Q \, \not{I} \, a_3$, Theorem 11 would imply $Q = O$; contradiction.

The case (11) is the dual of the preceding one. From now on we may assume that both the O_λ and the a_λ are mutually distinct.

If, for example, O_1, O_2, O_3 are not collinear, then $b = [O_2, O_3] \, \not{I} \, O_1$. As b is a fixed line of π_2 and π_3, we also have $b\pi_1 = b$. Hence $b = a_1$, by Theorem 11. Symmetrically, we obtain $a_2 = [O_3, O_1]$ and $a_3 = [O_1, O_2]$.

It remains to exclude the case that all the O_λ are incident with a line a and all the a_λ are incident with a point O. Choose

$$P \, I \, a_1, \quad P \, \not{I} \, a, \quad P \neq O.$$

Since $P = P\pi_1$, we also have $P = P\pi_2\pi_3$. Hence

$$P\pi_2 = P\pi_3^{-1} = Q, \quad \text{say}.$$

By (6), the points P, Q, O_2 and P, Q, O_3 are collinear. Since $P \, \not{I} \, a$ and $P \neq O$, we have $P \neq O_2$ and $P \, \not{I} \, a_2$. Hence $P \neq Q$ by Theorem 11. This yields O_2, $O_3 \, I \, [P, Q]$ and thus $[P, Q] = [O_2 \, O_3] = a$, in particular $P \, I \, a$; contradiction. □

We note the following

COROLLARY 14.1 *If the product of two perspective collineations is a perspective collineation, they have either the same centre or the same axis or the centre of either of them is incident with the axis of the other.*

We next discuss the commutativity of perspective collineations.

THEOREM 15 *Let* $\pi \in P(O, a)$, $\pi' \in P(O', a')$; $\pi \neq \iota \neq \pi'$; $\pi\pi' = \pi'\pi$.

7.5 THE CANONICAL DUALITY

Then either

$$a = a' \quad \text{and} \quad O = O' \tag{13}$$

or

$$a \mathbin{\mathrm{I}} O' \quad \text{and} \quad a' \mathbin{\mathrm{I}} O \tag{14}$$

or both.

PROOF Since $\pi' = \pi^{-1}\pi'\pi$, Theorem 12 implies $\pi' \in P(O'\pi, a'\pi)$. Hence $O'\pi = O'$ and $a'\pi = a'$. Thus by Theorem 11, either $O' = O$ or $O' \mathbin{\mathrm{I}} a$; also $a' = a$ or $a' \mathbin{\mathrm{I}} O$. Similarly, $\pi = \pi'^{-1}\pi\pi'$ implies that either $O = O'$ or $O \mathbin{\mathrm{I}} a'$, and either $a = a'$ or $a \mathbin{\mathrm{I}} O'$. This yields: either $O = O'$ or (14), and either $a = a'$ or (14). Altogether, we have either (13) or (14) [or both]. □

Note that (13) need not imply the commutativity of π and π' unless the projective plane is a Pappus plane. In a desarguesian projective plane the additional assumption $O \mathbin{\mathrm{I}} a$ is sufficient; cf. Chapter 8.

THEOREM 16 *Let* $\pi \in P(O, a)$, $\pi' \in P(O', a')$. *Assume* (14) *is satisfied but not* (13). *Then* $\pi\pi' = \pi'\pi$.

PROOF We may assume $a \neq a'$, the case $O \neq O'$ being its dual. Choose a point Q such that

$$Q \mathbin{\mathrm{I}} a, \quad Q \neq O, \quad Q \neq [a, a'].$$

Then $O' \mathbin{\mathrm{I}} a$ implies $Q\pi' \mathbin{\mathrm{I}} a$ and

$$Q\pi^{-1}\pi'\pi = Q\pi'\pi = Q\pi'. \tag{15}$$

By Theorem 12 and (14),

$$\pi^{-1}\pi'\pi \in P(O'\pi, a'\pi) = P(O', a').$$

Since π' also belongs to $P(O', a')$, (13) implies, by Theorem 17, $\pi^{-1}\pi'\pi = \pi'$ or $\pi\pi' = \pi'\pi$. □

7.5
THE CANONICAL DUALITY

Let $\mathfrak{P} = (\mathbf{P}, \mathbf{L}, \mathrm{I})$ denote any projective plane and let $\tilde{\mathfrak{P}}$ be the dual plane of $\mathfrak{P}$ obtained by interpreting each point of $\mathfrak{P}$ as a line of $\tilde{\mathfrak{P}}$ and each line of $\mathfrak{P}$ as a point of $\tilde{\mathfrak{P}}$. The incidence relation in $\mathfrak{P}$ is then interpreted as the incidence relation in $\tilde{\mathfrak{P}}$. Thus

$$\tilde{\mathfrak{P}} = (\tilde{\mathbf{P}}, \tilde{\mathbf{L}}, \mathrm{I}) \quad \text{where } \tilde{\mathbf{P}} = \mathbf{L}, \tilde{\mathbf{L}} = \mathbf{P}.$$

Let φ denote the pair of identical maps of $\mathbf{P}$ onto $\tilde{\mathbf{L}}$ and of $\mathbf{L}$ onto $\tilde{\mathbf{P}}$. Thus φ maps points onto lines and lines onto points preserving incidence:

$$P \operatorname{I} l \Leftrightarrow l\varphi \operatorname{I} P\varphi. \tag{16}$$

Thus the point P is incident with the line l in $\mathfrak{P}$ if and only if the point $l\varphi$ in $\tilde{\mathfrak{P}}$ is incident with the line $P\varphi$. We call φ the *canonical duality of* $\mathfrak{P}$ *onto* $\tilde{\mathfrak{P}}$ and write $\tilde{\mathfrak{P}} = \mathfrak{P}\varphi$.

If we denote the canonical duality of $\tilde{\mathfrak{P}}$ onto its dual plane $\tilde{\tilde{\mathfrak{P}}}$ by the same letter φ, we have

$$P\varphi^2 = P, \quad l\varphi^2 = l \quad \text{for every } P \in \mathbf{P}, l \in \mathbf{L}. \tag{17}$$

Thus $\tilde{\tilde{\mathfrak{P}}} = \mathfrak{P}$ and $\varphi^2 = \iota$.

If α is any collineation of $\mathfrak{P}$, $\varphi^{-1}\alpha\varphi$ will be one of $\tilde{\mathfrak{P}}$. We write

$$\alpha^\varphi = \varphi^{-1}\alpha\varphi.$$

If S is any set of collineations of $\mathfrak{P}$ put

$$S^\varphi = \{\alpha^\varphi \mid \alpha \in S\}.$$

For every choice of $O \in \mathbf{P}$ and $a \in \mathbf{L}$, we have

$$P(O)^\varphi = P(O\varphi), \quad P(a)^\varphi = P(a\varphi), \quad P(O, a)^\varphi = P(a\varphi, O\varphi) \tag{18}$$

and

$$P(a, b)^\varphi = P(b\varphi, a\varphi), \quad P(A, B)^\varphi = P(B\varphi, A\varphi). \tag{19}$$

We readily verify the following

LEMMA 17 *With $P(O, a)$, $P(a\varphi, O\varphi)$ is linearly transitive.*

PROOF Let P, Q denote two points collinear with and distinct from $a\varphi$; $P, Q \not\operatorname{I} O\varphi$. Then the lines $P\varphi^{-1}$ and $Q\varphi^{-1}$ are co-punctal with and distinct from a; $O \not\operatorname{I} P\varphi^{-1}, Q\varphi^{-1}$. Choose any line l through O not incident with $R = [a, P\varphi^{-1}]$ and put $S = [l, P\varphi^{-1}]$, $T = [l, Q\varphi^{-1}]$. By our assumption, there is a collineation $\pi \in P(O, a)$ such that $S\pi = T$. Hence

$$P\varphi^{-1}\pi = [R, S]\pi = [R, S\pi] = [R, T] = Q\varphi^{-1}$$

and thus $P\varphi^{-1}\pi\varphi = Q\varphi^{-1}\varphi = Q$, where $\varphi^{-1}\pi\varphi \in P(a\varphi, O\varphi)$. □

7.6
AFFINE RESTRICTIONS OF COLLINEATIONS OF PROJECTIVE PLANES

Let π denote any collineation of the projective plane $\mathfrak{P} = (\mathbf{P}, \mathbf{L}, \operatorname{I})$ and let l be a fixed line of π. By designating l as the improper line, we obtained an affine plane $\mathfrak{P}_l = (\mathbf{P}_l, \mathbf{L}_l, \operatorname{I}_l)$.

The restriction of π to $\mathbf{P}_l$ and $\mathbf{L}_l$ will be denoted by π_l. It maps $\mathbf{P}_l$ and $\mathbf{L}_l$ onto

7.6 AFFINE RESTRICTIONS OF COLLINEATIONS OF PROJECTIVE PLANES

themselves. Obviously, π_l is a collineation of the affine plane $\mathfrak{P}_l$. We call it the affine *restriction* of π to $\mathfrak{P}_l$. We have proved

THEOREM 18 *Let π be a collineation of the projective plane $\mathfrak{P}$ with the fixed line l. Then the restriction π_l of π to the affine plane $\mathfrak{P}_l$ is a collineation of $\mathfrak{P}_l$.*

EXAMPLES

1 Let $\pi \in P(O, a)$. Since any point on a is a fixed point of π, any parallel pencil Π is mapped onto itself by π_a. Thus $h\pi_a \parallel h$ for every line h in $\mathfrak{P}_a$ and π_a is a homothety.

If $O \not\!I\, a$, the proper point O is a fixed point of π_a and π_a is a dilatation with the centre O.

If $O\,\mathrm{I}\,a$, the pencil of the lines $\neq a$ through O becomes a parallel pencil Π_O in $\mathfrak{P}_a$. As these lines are the traces of π and π_a, π_a becomes a translation parallel to Π_O.

2 Assume again $\pi \in P(O, a)$. Let l be a line through O distinct from a. Since all the proper points on the proper line a are fixed points, π_l is an affinity with the axis a.

THEOREM 19 *Let $\mathfrak{A} = (\mathbf{P}, \mathbf{L}, \mathrm{I})$ be any affine plane. Then every collineation α of $\mathfrak{A}$ is the restriction of one and only one collineation α^* of the projective closure*

$$\mathfrak{A}^* = (\mathbf{P^*}, \mathbf{L^*}, \mathrm{I}^*) \text{ of } \mathfrak{A}.$$

We call α^* the *projective extension* of α.

PROOF We first verify the uniqueness of the projective extension of α. Let α^*, α^0 be two projective extensions of α. Then $h\alpha^* = h\alpha = h\alpha^2$ for all $h \in \mathbf{L}$. Thus the collineation $\beta = \alpha^{*-1}\alpha^0$ of $\mathfrak{A}^*$ satisfies

$$h\beta = h \quad \text{for every } h \text{ in } \mathbf{L^*}.$$

Every point P is the intersection of two distinct lines in $\mathbf{L^*}$. As they are fixed, P remains fixed. Hence $\beta = \iota$ and $\alpha^* = \alpha^0$.

We next prove the existence of a projective extension of α. By Chapter 2, Corollary 2.1, α maps a parallel pencil onto a parallel pencil. Thus the equations

$$\begin{cases} P\alpha^* = P\alpha & \text{for } P \in \mathbf{P}, \\ h\alpha^* = h\alpha & \text{for } h \in \mathbf{L}, \\ P_\omega\alpha^* = Q_\omega & \text{such that } P_\omega = \Pi \text{ implies } Q_\omega = \Pi\alpha \text{ (cf. p. 85),} \\ l_\omega\alpha^* = l_\omega \end{cases} \quad (20)$$

define a pair α^* of maps of $\mathbf{P^*}$ and $\mathbf{L^*}$ into themselves. Their restrictions to P and L is the pair of bijections α of $\mathbf{P}$ and $\mathbf{L}$ onto themselves.

By Chapter 2, Corollary 2.2, α maps the set of all the parallel pencils bijectively onto itself. Thus the restriction of α^* to the set of the improper points is a bijection. Hence $\alpha^*: \mathbf{P}^* \to \mathbf{P}^*$ is a bijection.

Since the restriction of α^* to $\mathbf{L}$ is bijective and $\mathbf{L}^* = \{l_\omega\} \cup \mathbf{L}$, the equation $l_\omega \alpha^* = l_\omega$ implies that $\alpha^*: \mathbf{L}^* \to \mathbf{L}^*$ is also bijective.

It remains to verify that α^* preserves incidence. Let

$$P \mathbf{I}^* h, \qquad P \in \mathbf{P}^*, \qquad h \in \mathbf{L}^*.$$

We have to show that $P\alpha^* \mathbf{I}^* h\alpha^*$. Since $P \notin \mathbf{P}_\omega$ and $P \mathbf{I}^* h$ implies $h \neq l_\omega$, there are only three cases:

(i) $P \in \mathbf{P}$ and $h \in \mathbf{L}$. Then by the definition of α, $P\alpha \mathbf{I} h\alpha$ and thus $P\alpha^* \mathbf{I}^* h\alpha^*$.

(ii) $P \notin \mathbf{P}$ and $h \in \mathbf{L}$. Then $P = P_\omega$ and $P\alpha^* = P_\omega \alpha^* = Q_\omega$ if $P_\omega = \Pi$ and $Q_\omega = \Pi\alpha$. From $P \mathbf{I}^* h$ it follows that $h \in \Pi$; thus

$$h\alpha \in \Pi\alpha \text{ and } P\alpha^* = Q_\omega \mathbf{I}^* h\alpha = h\alpha^*.$$

(iii) $P \notin \mathbf{P}$ and $h \notin \mathbf{L}$. Then $P = P_\omega$ and $h = l_\omega$. Hence $P\alpha^* = P_\omega \alpha^* = Q_\omega \in \mathbf{P}_\omega$ and $h\alpha^* = l_\omega \alpha^* = l_\omega = h$; thus $P\alpha^* \mathbf{I}^* h\alpha^*$.

We have shown that α^* is a collineation of $\mathfrak{A}^*$. By (20), the restriction α^*_l of α^* is equal to α if $l = l_\omega$. $\square$

EXAMPLES Let $\mathfrak{A}^* = (\mathbf{P}^*, \mathbf{L}^*, \mathbf{I}^*)$ denote the projective closure of the affine plane $\mathfrak{A} = (\mathbf{P}, \mathbf{L}, \mathbf{I})$.

1 Let $\alpha \in H$. Then α maps every parallel pencil onto itself. Hence every point of l_ω is a fixed point of α^*. Thus α^* has the axis l_ω.

If $\alpha \in D(O)$, O is a centre both of α and of α^*. Thus $\alpha^* \in P(O, l_\omega)$.

If $\alpha \in T(\Pi)$, all the lines of Π are fixed lines of α and of α^*, and $P_\omega = \Pi$ is a centre of α^*; cf. (20). Thus $\alpha^* \in P(P_\omega, l_\omega)$.

2 Let $\alpha \in A(a)$. As all the points on a are fixed points of α, a is an axis of α^*. The traces of α form a parallel pencil Π. Hence $\alpha^* \in P(P_\omega, a)$ if $\Pi = P_\omega$.

COROLLARY 19.1 *Let π be a collineation of the projective plane $\mathfrak{P}$; $l\pi = l$. Then if $(\mathfrak{P}_l)^*$ is identified with $\mathfrak{P}$,*

$$(\pi_l)^* = \pi. \tag{21}$$

PROOF Both sides of (21) have the same restriction π_l in $\mathfrak{P}_l$; cf. Theorem 19. $\square$

The reader will verify

THEOREM 20. *Let $\mathfrak{A}$ be any affine plane with the projective closure $\mathfrak{A}^*$ and the improper line $l = l_\omega$. If every collineation α of $\mathfrak{A}$ is mapped onto its projective extension, we obtain an isomorphism of the group of the collineations of $\mathfrak{A}$ onto the group of those collineations of $\mathfrak{A}^*$ which have the fixed line l.*

The isomorphism of Theorem 20 induces an isomorphism of each subgroup of the group of the collineations of $\mathfrak{A}$ onto some subgroup of the group of those of $\mathfrak{A}^*$. In particular, the examples of this section yield the following isomorphisms:

$H \cong P(l)$, $D(O) \cong P(O, l)$ if $O \not{I} l$;

$T(\Pi_o) \cong P(O, l)$ if $O I l$.

Also

$T \cong P(l, l)$, $A(a) \cong P(l, a)$ $l \neq a$;

$H(b) \cong P(b, l)$, $b \neq l$.

Conversely, we may start out with a group of collineations π in a projective plane $\mathfrak{P}$ with the common fixed line l and construct the group of their restrictions π_l in $\mathfrak{P}_l$; cf. Theorem 18 and the examples on pp. 97 and 98. We then readily obtain

THEOREM 21

$$\{\pi_l \mid \pi \in P(l)\} = H \tag{22}$$

[= group of the homotheties of $\mathfrak{P}_l$];

$$\{\pi_l \mid \pi \in P(O, l)\} = \begin{cases} D(O) & \text{if } O \not{I} l, \\ T(\Pi_o) & \text{if } O I l; \end{cases} \tag{23}$$

$$\{\pi_l \mid \pi \in P(a, b)\} = \begin{cases} H_a & \text{if } b = l \neq a, \\ T & \text{if } a = b = l, \\ A(b) & \text{if } a = l \neq b. \end{cases} \tag{24}$$

Every homothety or axial affinity in $\mathfrak{A}$ is the affine restriction of a perspective collineation in $\mathfrak{A}^*$. Conversely, the affine restriction of any perspective collineation of $\mathfrak{A}^*$ with an improper axis or centre is either a homothety or an axial affinity. Therefore many theorems in Chapter 2 are special cases of theorems of this chapter.

EXERCISES

1 Consider the following:

Axiom P4
(i) Given any two distinct points there is one and only one line incident with both.
(ii) To any two distinct lines there is one and only one point incident with both.
(i) Show that P4 ⇔ P4(i) ∧ P1(ii). (ii) Show that P1 ∧ P2 ⇔ P4. (iii) Prove: The set of axioms P1–P3 is equivalent to the pair of axioms P3 ∧ P4.

2 Prove Theorem 3.

3 Verify the statements of Example 3.1, p. 88.

4 A projective plane is *finite* if either the number of points or that of lines is finite. Prove: If the projective plane $\mathfrak{P}$ is finite, there exists a number n, the *order* of $\mathfrak{P}$, such that every line is incident with exactly $n+1$ points and every point with exactly $n+1$ lines. $\mathfrak{P}$ then has exactly n^2+n+1 points and lines, each. Find a weakest possible definition of finite projective planes.

5 Every trace of an involutory collineation is fixed.

6 Suppose every trace of the collineation $\gamma \neq \iota$ is fixed.
(i) Show that every dual trace is fixed.
(ii) Show that γ is either perspective or *planar* [i.e. the set $\mathfrak{P}'$ of the fixed elements of γ is a *sub-plane* of $\mathfrak{P}$. Thus $\mathfrak{P}'$ satisfies the axioms P1–P3].
(iii) Assume in addition that $\mathfrak{P}$ is finite, say of order n, and that γ is planar. Then n is a square and the order of $\mathfrak{P}'$ is equal to $\sqrt{n}$. Also there is an integer m such that for every point P and every line l, neither in $\mathfrak{P}'$,
$$P\gamma^k = P \leftrightarrow l\gamma^k = l \leftrightarrow \gamma^k = \iota \leftrightarrow m|k.$$

7* Let $a, b \in \mathbf{L}$, $a \neq b$. Suppose $P(a, b)$ is linearly transitive and has an automorphism ϕ which maps $P(O, b)$ onto itself for each $O \mathrel{\mathrm{I}} a$. Let $A, B \mathrel{\mathrm{I}} a, b$. Show that there is a collineation α with $A\alpha = B$ such that
$$\tau^\phi = \alpha^{-1}\tau\alpha \quad \text{for every } \tau \in P(a, b).$$
The group $P(b, a)$ is linearly transitive, and ϕ is the identity; cf. the proof of Chapter 4, Theorem 7.

8 Let $O \mathrel{\mathrm{I}} l$. Then the group $\{\pi_l | \pi \in P(O, O)\}$ is the union of $T(\Pi_O)$ and the set of the shears parallel to Π_O. Prove once more that this group is abelian; cf. Chapter 4, Exercise 6.

9 Let $a \notin \Pi$. A *reflection in a parallel to* Π is an involutory affinity with the axis a and the pencil of traces Π. Applying a duality to the projective closure, state and prove the analogue of Chapter 2, Theorem 24 for these reflections.

10 Let $\mathfrak{A} = (\mathbf{P}, \mathbf{L}, \mathrel{\mathrm{I}})$ be any affine plane. Omitting either from $\mathbf{L}$ all the lines of a parallel pencil or one point from $\mathbf{P}$ and the lines through that point from $\mathbf{L}$, we obtain an *incomplete affine plane*. Establish sets of axioms for both types of incomplete affine planes and prove that an incomplete affine plane defined by means of these axioms can be completed to an affine plane in one and only one way. Show that both types of incomplete affine planes satisfy a duality principle. State and prove the analogue of Theorem 20 for these planes.

8
Desarguesian projective planes

8.1
PROJECTIVE AND AFFINE DESARGUESIAN PLANES

We call the projective plane $\mathfrak{P}$ *desarguesian* if all the groups $P(O, a)$ are linearly transitive. Thus in every affine restriction $\mathfrak{P}_l$, all the groups $D(O)$ are linearly transitive and $\mathfrak{P}_l$ is desarguesian.

Conversely, assume that every $\mathfrak{P}_l$ is desarguesian. By Chapter 4, Corollary 2.1, $\mathfrak{P}_l$ is a translation plane. Thus in every $\mathfrak{P}_l$, not only all the groups $D(O)$ but also each $T(\Pi)$ is linearly transitive; i.e. all the groups $P(O, l)$ in $\mathfrak{P}$ are linearly transitive. As this applies to every choice of l, $\mathfrak{P}$ is desarguesian. This yields

THEOREM 1 *The plane $\mathfrak{P}$ is desarguesian if and only if every restriction of $\mathfrak{P}$ is desarguesian.*

In order to improve the preceding theorem we use the dual $\bar{\mathfrak{P}}$ of $\mathfrak{P}$. Section 7.5 implies

LEMMA 2.1 *A projective plane is desarguesian if and only if its dual plane is.*

We can now state

THEOREM 2 *A projective plane is desarguesian if one affine restriction is.*

PROOF Suppose $\mathfrak{P}_l$ is desarguesian. Then by Chapter 4, Theorem 6 and by 7.6, we have:

(A) Every group $P(O, a)$ with $a = l$ or $O\,\mathrm{I}\,l$ is linearly transitive in $\mathfrak{P}$.

We have to prove the linear transitivity of $P(O, a)$ for every choice of O and P.

Let φ denote the canonical duality of $\mathfrak{P}$ onto its dual plane $\bar{\mathfrak{P}}$. Let $L = l\varphi$. By Chapter 7, Lemma 17, (A) yields

(B) Every group $P(\bar{O}, \bar{a})$ in $\bar{\mathfrak{P}}$ with $\bar{O} = L$ or $\bar{a}\,\mathrm{I}\,L$ is linearly transitive.

Choose any line h through L. For every $\bar{O}\,\mathrm{I}\,h$, the group $P(\bar{O}, h)$ is linearly transitive by (B). Hence its restriction $D(\bar{O})$ to $\bar{\mathfrak{P}}_h$ is linearly transitive too. This implies that $\bar{\mathfrak{P}}_h$ is desarguesian. Applying Chapter 4, Theorem 6 and 7.5 once more, we obtain:

(C) Let $h\,\mathrm{I}\,L$. Then the group $P(h, \bar{a})$ is linearly transitive in $\bar{\mathfrak{P}}$ for every $\bar{a} \in L$.

Since every point $\bar{O}$ of $\mathfrak{P}$ is on some line h through L, (C) implies that the group $P(\bar{O}, \bar{a})$ is linearly transitive for every choice of $\bar{O}$ and $\bar{a}$ in $\mathfrak{P}$, i.e., that $\mathfrak{P}$ is desarguesian. Hence by Lemma 2.1, $\mathfrak{P}$ is desarguesian. □

Theorem 2 can be restated as follows:

THEOREM 2′ *The projective closure of an affine desarguesian plane is desarguesian.*

8.2
THE PROJECTIVE THEOREM OF DESARGUES

In Chapter 4, the affine restriction $\mathfrak{P}_l$ of $\mathfrak{P}$ was defined to be desarguesian if and only if the affine theorem of Desargues held true in $\mathfrak{P}_l$. In order to formulate the following theorem more conveniently, we define: let $P_0 P_1 P_2$ and $Q_0 Q_1 Q_2$ denote two triangles in $\mathfrak{P}$ with the sides

$$p_0 = [P_1, P_2], \quad p_1 = [P_2, P_0], \quad p_2 = [P_0, P_1]$$

and

$$q_0 = [Q_1, Q_2], \quad q_1 = [Q_2, Q_0], \quad q_2 = [Q_0, Q_1],$$

respectively. They are called *perspective from the centre* O if the points O, P_k, Q_k are collinear; they are *perspective from the axis* a if the lines a, p_k, q_k are copunctal; $k = 0, 1, 2$.

STATEMENT (D)* ['Projective Theorem of Desargues']. *If two triangles are perspective from a centre, they are also perspective from an axis*; cf. Figure 8.1.

THEOREM 3 *The plane $\mathfrak{P}$ is desarguesian if and only if Statement (D)* holds true in $\mathfrak{P}$.*

PROOF If (D)* holds true in $\mathfrak{P}$, then (D) will be valid in every restriction $\mathfrak{P}_a$ of $\mathfrak{P}$. Thus every $\mathfrak{P}_a$ is desarguesian and so is $\mathfrak{P}$ by Theorem 1.

Conversely, let $\mathfrak{P}$ be desarguesian and let $P_0 P_1 P_2$ and $Q_0 Q_1 Q_2$ be perspective from O. If, for example, $p_0 = q_0$, choose a point incident with both p_k and q_k; $k = 1, 2$. Then the triangles are perspective from a line through these two points. Thus we may assume that $p_k \neq q_k$ for $k = 0, 1, 2$. We have to show that the three points

$$R_k = [p_k, q_k] \quad (k = 0, 1, 2)$$

are collinear. We may assume that they are mutually distinct. Then none of the points P_k, Q_k is incident with $a = [R_1, R_2]$. Designate a as improper. In $\mathfrak{P}_a$, the triangles $P_0 P_1 P_2$ and $Q_0 Q_1 Q_2$ satisfy the assumptions of Statement (D). Thus we have $p_1 \parallel q_1$ and $p_2 \parallel q_2$ in $\mathfrak{P}_a$. Since $\mathfrak{P}_a$ is desarguesian, this implies $p_0 \parallel q_0$ in $\mathfrak{P}_a$, i.e., $R_0 \, \mathrm{I} \, a$ in $\mathfrak{P}$. □

8.2 THE PROJECTIVE THEOREM OF DESARGUES

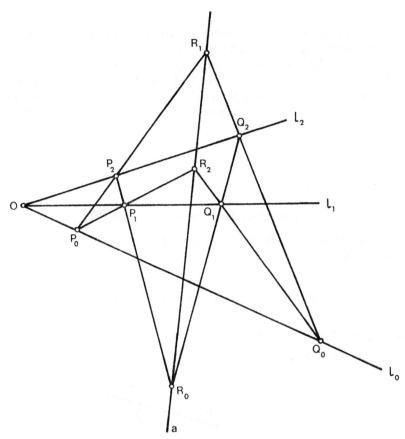

Figure 8.1

Combining Theorem 3 with Lemma 2.1 we obtain

COROLLARY 3.1 *Statement (D)* is equivalent to its converse: If two triangles are perspective from an axis they are also perspective from a centre.*

The two triangles $P_0 P_1 P_2$ and $Q_0 Q_1 Q_2$ perspective from the centre O and the axis a, together with the points R_k and the lines p_k, q_k and l_k form a *Desargues configuration*. We do not exclude the possibility that some of the ten points or ten lines of this configuration are identical or that there are additional incidences. Thus the case $O\,\mathrm{I}\,a$ yields a projective generalization of the little affine theorem (d) of Desargues.

Let $(D)^*_{O,a}$ denote Statement $(D)^*$ with the additional requirement that the centre and axis are equal to a given point O and line a, respectively.

THEOREM 4 *The group $P(O, a)$ is linearly transitive if and only if Statement $(D)^*_{O,a}$ holds true in $\mathfrak{P}$.*

PROOF

(i) Let $O \not{I} a$. If we designate a as improper, the group $P(O, a)$ will be restricted to the group $D(O)$ and Statement (D)*$_{O,a}$ becomes Statement (D)$_O$. Thus Chapter 4, Corollary 3.1, implies our assertion in this case.

(ii) Let $O I a$. We designate a as improper and put $O = P_\Pi$. Then the group $P(O, a)$ and Statement (D)*$_{O,a}$ correspond to the group $T(\Pi)$ and the Statement (d)$_\Pi$ respectively. Hence our assertion now follows from Chapter 3, Corollary 5.1. □

8.3
PROJECTIVE PAPPUS PLANES

It remains to discuss the projective extension of Statement (P).

STATEMENT (P)* ['Projective Theorem of Pappus']. *Suppose no two consecutive points of the cyclic sequence*

$$\ldots A_1, B_2, C_1, A_2, B_1, C_2, A_1, \ldots$$

are collinear with any of the remaining ones, but the three points A_k, B_k, C_k are collinear; $k = 1, 2$. Then the three points

$$A_0 = [[B_1, C_2], [C_1, B_2]], \quad B_0 = [[C_1, A_2], [A_1, C_2]], \quad C_0 = [[A_1, B_2], [B_1, A_2]]$$

are collinear; cf. Figure 8.2.

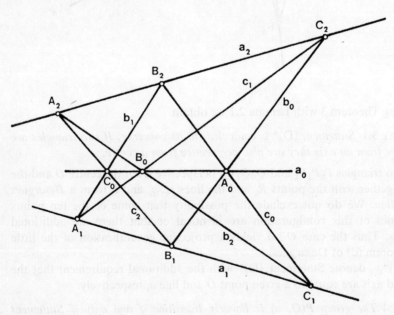

Figure 8.2

We call the projective plane $\mathfrak{P}$ a *Pappus plane* if Statement (P)* holds true. Thus $\mathfrak{P}$ is a Pappus plane if every affine restriction of $\mathfrak{P}$ is one. By Chapter 5, Corollary 2.1, this is equivalent to every group $P(O, a)$ being linearly transitive and abelian.

THEOREM 5 *The projective closure of an affine Pappus plane is a Pappus plane.*

PROOF Let $\mathfrak{A}$ be an affine Pappus plane. Then $\mathfrak{A}$ is in particular desarguesian. By Theorem 2', $\mathfrak{A}^*$ is desarguesian too; thus all the groups $P(O, a)$ in $\mathfrak{A}^*$ are linearly transitive. Hence to any pair O, a there is a collineation which maps l onto a. Let Q denote the inverse image of O. Since the groups $P(Q, l)$ and $P(O, a)$ are conjugate and the former is abelian, $P(O, a)$ must also be abelian. □

Let $\mathfrak{P}$ be a Pappus plane and let φ denote the canonical duality of $\mathfrak{P}$. By 7.5, the groups $P(a\varphi, O\varphi)$ and $P(O, a)$ are isomorphic. In particular, if one of them is abelian, so is the other. Hence Lemma 2.1 yields:

REMARK 5.1 *A projective plane is a Pappus plane if and only if its dual is.*

The dual of the projective theorem of Pappus is the following statement:
'Theorem of Pappus-Brianchon.' *Suppose no two consecutive lines of the cyclic sequence*

$$\ldots a_1, b_2, c_1, a_2, b_1, c_2, a_1, \ldots$$

are co-punctal with any of the remaining ones, but the three lines a_k, b_k, c_k are co-punctal; $k = 1, 2$. Then the three lines

$$a_0 = [[b_1, c_2], [c_1, b_2]], \quad b_0 = [[c_1, a_2], [a_1, c_2]], \quad c_0 = [[a_1, b_2], [b_1, a_2]]$$

are co-punctal.

THEOREM 6 $\mathfrak{P}$ *is a Pappus plane if and only if the theorem of Pappus-Brianchon holds true in* $\mathfrak{P}$.

PROOF The theorem of Pappus-Brianchon is valid in $\mathfrak{P}$ if and only if the theorem of Pappus holds in the dual plane. Thus Remark 5.1 implies our assertion. □

Note that the theorem of Pappus-Brianchon is identical with the configuration of that of the theorem of Pappus; cf. Figure 8.2. Thus it can also be proved without the canonical duality.

EXERCISES

1 Transfer Chapter 4, Exercise 1 to projective planes.

2 Suppose no three of the four points P_1, P_2, P_3, P_4 and Q_1, Q_2, Q_3, Q_4, respectively, of a desarguesian projective plane are collinear. Construct a collineation π such that $P_k\pi = Q_k$ for $k = 1, 2, 3, 4$.

3 Let $(\overline{D})$ denote the statement obtained from (D)* by designating a line through the centre as improper. Show that $(\overline{D})$ holds in an affine plane if and only if it is desarguesian.

4 Suppose that for every pair $O \mathrel{I} a$ the group $P(O, a)$ contains at least two elements. Then all the elations $\neq \iota$ have the same order. If it is finite it is a prime number.

5 Suppose the three lines l, l', l'' in the projective plane $\mathfrak{P}$ are not co-punctal and the three affine restrictions $\mathfrak{P}_l, \mathfrak{P}_{l'}, \mathfrak{P}_{l''}$ of $\mathfrak{P}$ are translation planes. Show that every affine restriction of $\mathfrak{P}$ is a translation plane.

6* Construct a finite projective plane $\mathfrak{P}$ which possesses two lines l and l' such that $\mathfrak{P}_l$ is a translation plane while $\mathfrak{P}_{l'}$ is not; cf. Chapter 4, Exercise 7.

Appendix

A1
GROUPS

The non-void set G with the elements $a, b, \ldots, x, y, \ldots$ is a *group* if there is a mapping $(x, y) \mapsto xy$ of $G \times G$ onto G satisfying the following conditions:
(i) $x(yz) = (xy)z$;
(ii) there is one and only one element $e \in G$ such that $ex = xe = x$ for all $x \in G$;
(iii) for each $x \in G$, there is precisely one element $x' \in G$ such that $xx' = x'x = e$. We write $x' = x^{-1}$.
If in addition
(iv) $xy = yx$ for all x, y in G,

the group G is *abelian*. Abelian groups are also written additively. In that case, the element x' of (iii) will be written $x' = -x$.

Let H be a subset of G. Then H is a *subgroup* of G if H is a group if the composition in H is defined as in G. It is readily seen that the subset H of G is a subgroup of G if and only if $x \in H$, $y \in H$ implies $xy^{-1} \in H$.

An *isomorphism* of the group G onto the group $\tilde{G}$ is a bijection $\varphi: G \to \tilde{G}$ such that

$$\varphi(xy) = \varphi(x)\varphi(y) \qquad \text{for all } x, y \text{ in } G.$$

It maps the unit element e of G onto the unit element $\tilde{e}$ of $\tilde{G}$ and inverses onto inverses:

$$\varphi(e) = \tilde{e} \text{ and } \varphi(x^{-1}) = (\varphi(x))^{-1} \qquad \text{for all } x \in G.$$

An *automorphism* of G is an isomorphism of G onto itself. An important class of automorphisms are the *inner automorphisms*: Let $a \in G$. Then

$$x \mapsto x^a = a^{-1}xa$$

is readily seen to define an automorphism of G. It is called the inner automorphism induced by the element a. For any x and any a in G, the elements x and x^a are called *conjugate*. Conjugacy is an equivalence relation. If H is a subgroup of G, the subgroups H and $H^a = \{x^a \mid x \in H\}$ are *conjugate*. If a subgroup is identical with all of its conjugates, it is *normal*. Thus a subgroup N is normal if and only if

$$a^{-1}Na = N \quad \text{for all } a \in G.$$

A2
SKEW FIELDS

A set F with the elements $x, y, \ldots$ is a *skew field* if two operations, addition and multiplication, are defined in F with the following properties:
(i) F is an abelian group under addition. Let 0 denote the neutral element of this group.
(ii) The set $F^* = F \setminus \{0\}$ is a multiplicative group. We denote its neutral element by 1.
(iii) Addition and multiplication are tied by the distributive laws:

$$x(y+z) = xy+xz, \quad (y+z)x = yx+zx \quad \text{for all } x, y, z \text{ in } F.$$

By (iii), $x \cdot 0 = 0 \cdot x = 0$ for every x. Hence (ii) readily implies that the multiplication in F is associative.

If the multiplication is commutative, F is a *field*. By a theorem of Wedderburn, every finite skew field must be a field. The simplest example of a non-commutative skew field is the quaternion field $\mathbb{H}$ over the field $\mathbb{R}$ of the real numbers; cf. Chapter 6, Exercise 1.

A bijective map α of F onto a second skew field (onto itself) is an isomorphism (automorphism) if

$$(x+y)\alpha = x\alpha+y\alpha \quad \text{and} \quad (xy)\alpha = x\alpha \cdot y\alpha \quad \text{for all } x, y \text{ in } F.$$

A3
RIGHT VECTOR SPACES

The set $\mathbf{V}$ with the elements $A, B, \ldots$ is a *right vector space* over the skew field F if it satisfies the following conditions:
(i) $\mathbf{V}$ is an additive abelian group. Its neutral element is the *null-vector* O.
(ii) Every pair $x \in F$, $A \in \mathbf{V}$ determines a vector Ax such that

$$A \cdot 1 = A, \qquad A(xy) = (Ax)y,$$
$$(A+B)x = Ax+Bx, \qquad A(x+y) = Ax+Ay$$

for all x, y in F and all A, B in $\mathbf{V}$.

The elements of a (right) vector space are called *vectors*.

A non-void subset of **V** which is closed under these two operations is again a right vector space over F. It is called a *subspace* of **V**.

Given a finite subset

$$\{A_1, \ldots, A_k\} \tag{1}$$

of **V**, the set

$$\langle A_1, \ldots, A_k \rangle$$

of all the *linear combinations*

$$A_1 x_1 + \ldots + A_k x_k$$

is a subspace of **V**, the subspace *spanned* by the elements of (1). It is the intersection of all the subspaces of **V** which contain the set (1). In particular

$$\langle A \rangle = \{Ax \mid x \in F\}.$$

If A and B are both distinct from O, then

$$\langle A \rangle = \langle B \rangle \Leftrightarrow \langle A \rangle \cap \langle B \rangle \neq \{O\} \Leftrightarrow A \text{ is a multiple of } B. \tag{2}$$

REMARK Given the subspace $\langle A \rangle$ and the vector B, the set

$$B + \langle A \rangle = \{B + X \mid X \in \langle A \rangle\}$$

is a *coset* of $\langle A \rangle$ in **V**. By (2), any two cosets are either disjoint or identical.

The vectors $A_1, \ldots, A_k$ are *linearly independent* if the equation

$$A_1 x_1 + \ldots + A_k x_k = O$$

has only the *trivial solution*

$$x_1 = \ldots = x_k = 0.$$

Otherwise they are *linearly dependent*.

If the vector space **V** has n linearly independent vectors which span **V**, the set of these vectors is a *base* of **V**. Then each vector of **V** can be expressed as a linear combination of these vectors in one and only one way and any other base also consists of precisely n elements. The number n is the *dimension* of **V** and denoted by dim **V**.

If dim **V** $= n$ and **V**' is a subspace of **V**, **V**' has a dimension. We have dim **V**' $\leq$ dim **V**, equality holding only if **V**' $=$ **V**.

Let **V** and **V**' denote right vector spaces over the skew fields F and F', respectively. If the isomorphism μ of F onto F' and the bijective map φ of **V** onto **V**' satisfy

$$(A+B)\varphi = A\varphi + B\varphi, \qquad (Ax)\varphi = (A\varphi) \cdot x^\mu$$

for all A, B in **V** and all x in F, the pair (φ, μ) is called an *isomorphism* of **V** onto

V'. It is *linear* if $F = F'$ and μ is the identity. Isomorphisms map subspaces onto subspaces and preserve dimensions.

If $F = F'$ and $\mathbf{V} = \mathbf{V}'$, (φ, μ) is called a *[semi-linear] automorphism* of **V**. The automorphisms of **V** form a group $\Gamma(\mathbf{V})$, if the product of the automorphisms (φ, μ) and (φ', μ') is the automorphism $(\varphi\varphi', \mu\mu')$. The linear automorphisms of **V** form a normal subgroup of $\Gamma(\mathbf{V})$.

The set

$$F^n = \{(a_1, \ldots, a_n) \mid a_1, \ldots, a_n \text{ in } F\}$$

of all the ordered n-tuplets of elements of F becomes an n-dimensional right vector space over F if addition and multiplication by *scalars*, i.e., by elements of F, are defined through

$$(x_n, \ldots, x_n) + (y_1, \ldots, y_n) = (x_1 + y_1, \ldots, x_n + y_n),$$

$$(x_1, \ldots, x_n)z = (x_1 z \ldots, x_n z).$$

The vectors

$$E_1 = (1, 0, \ldots, 0), \quad E_2 = (0, 1, 0, \ldots, 0), \quad \ldots, \quad E_n = (0, \ldots, 0, 1)$$

form a base of F^n.

If **V** is any n-dimensional vector space over F with the base $A_1, \ldots, A_n$, there is a linear isomorphism φ of **V** onto F^n such that $A_k \varphi = E_k$; $k = 1, \ldots, n$. It satisfies

$$\left(\sum_{1}^{n} A_h x_h\right)\varphi = \sum E_h x_h = (x_1, \ldots, x_n).$$

References

F. Bachmann. *Aufbau der Geometrie aus dem Spiegelungsbegriff;* second edition (Berlin, 1973)

R. Baer. *Linear algebra and projective geometry* (New York, 1952)

W. Benz. Über Möbiusebenen. Jahresber. D.M.V. *63* (1960), 1–27

W. Benz and H. Mäurer. Über die Grundlagen der Laguerre-Geometrie. Jahresber. D.M.V. *67* (1964), 14–42

R. J. Bumcrot. *Modern projective geometry* (New York, 1969)

H. S. M. Coxeter. *The real projective plane;* second edition (Cambridge, 1960)

D. Hilbert. *Grundlagen der Geometrie;* ninth edition (Stuttgart, 1962)

– *The foundations of geometry* (Chicago, 1902)

D. R. Hughes and F. C. Piper. *Projective planes* (New York, 1973)

H. Karzel. Bericht über projektive Inzidenzgruppen. Jahresber. D.M.V. *67* (1964), 58–92

H. Lenz. *Vorlesungen über projektive Geometrie* (Leipzig, 1965)

R. Lingenberg. *Grundlagen der Geometrie* I (Mannheim, 1969)

– Metrische Geometrie der Ebene und S-Gruppen, Jahresber. D.M.V. *69* (1966), 9–50

H. R. Salzmann. Topological planes, Advances in Math. *2*, fasc. 1. (1967), 1–60

O. Veblen and J. W. Young. *Projective geometry*, 2 vols (Boston, 1916–18)

Index

affine plane 6; of order two 11
affinity 83; axial 28
axis, of a collineation 20, 91

centre, of a collineation 20, 91
characteristic, of a translation plane 44
collineation 15, 90
collinear points 6
commuting collineations 19, 25, 29, 94, 95
co-ordinate planes 8, 73
co-punctal lines 89

Desargues 32; configuration 103; affine theorem of 52; little affine theorem of 32; projective theorem of 102
desarguesian planes 52, 101
dilatation 26
direction, of a translation 23
duality 90; principle of 90; canonical 96

elation 92
extension, projective, of an affine collineation 97

finite plane 11
fixed line 19, 91; point 19, 91; set 19

homology 92
homothety 21

incidence 3, 86
intersection, of lines 4; table 10
inverse, of a bijection 14; of a collineation 16

involution 27
isomorphism, of affine planes 15; of projective planes 91

midpoint 51

order, of a plane 11, 100

Pappus, 35; affine theorem of 66; little affine theorem of 35; projective theorem of 104; theorem of, and Brianchon 105; plane 66, 105
perspective collineation 92; triangles 102
planar collineation 100
prime kernel, of a translation plane 44
pencil 5, 6
product, of two collineations 16
projective plane 87
projective closure, of an affine plane 87

quaternions 84, 108

reflection 27, 100
restriction, affine, of a projective collineation 97

shear 64; theorem 57; little, theorem 36
subplane 19

trace, of a collineation 22, 91; dual 91
transitivity, linear 32, 93
translation 23; plane 32